Hatching & Brooding Your Own Chicks

Hatching & Brooding
YOUR OWN CHICKS

Chickens • Turkeys • Ducks • Geese • Guinea Fowl

Gail Damerow

Storey Publishing

The mission of Storey Publishing is to serve our customers by
publishing practical information that encourages
personal independence in harmony with the environment.

Edited by Deborah Burns and Rebekah Boyd-Owens
Art direction and book design by Mary Winkelman Velgos
Text production by Jennifer Jepson Smith

Cover photography by © Arco Martin/Getty Images: front, right; © fotolincs/Alamy: front, bottom left; © Fuse/Getty Images: front, center; © Gail Damerow: back, left two eggs; © Mark Mosrie: back, author; Mars Vilaubi: back, right two eggs; © Photoshot/Alamy: front, top left; © Photoshot/Ernie James: front, title
Interior photography credits appear on page 239
Illustrations by Bethany Caskey, except for chart, page 212, by Ilona Sherratt

Indexed by Samantha Miller

Storey Publishing
210 MASS MoCA Way
North Adams, MA 01247
www.storey.com

Printed in the United States by Quad/Graphics
10 9 8 7 6 5 4 3 2 1

Library of Congress Cataloging-in-Publication Data

Damerow, Gail.
  Hatching & brooding your own chicks / by Gail Damerow.
    p. cm.
  Includes index.
  ISBN 978-1-61212-014-0 (pbk. : alk. paper)
  1. Poultry. 2. Chicks. I. Title. II. Title: Hatching and brooding your own chicks.
SF495.D36 2013
636.5—dc23
                    2012013937

CONTENTS

INTRODUCTION

Which comes first, the chicken or the egg? I mulled over this conundrum while deciding how to organize this book. A certain logic seems to dictate starting with eggs and incubation. But that entails an awfully steep learning curve for the poultry-keeping newbie.

Because the more common scenario is to start a flock with chicks from the local farm store or ordered by mail, I chose to begin this book with details on acquiring and brooding chicks. By the time the pullets start laying, you are comfortable with brooding and raising chicks, and excited by the possibility of hatching some of their eggs.

Other folks start out with grown-up hens, as did I. When I bought my first house, I wanted to make sure it was zoned for chickens, and what better way than to buy a place that came with hens already installed! Come spring I was at the farm store, heard peeping, and couldn't help bringing home a boxful of fluffy chicks to raise. The folks at the store provided helpful information on how to brood chicks in a cardboard box using a light bulb for heat. What they didn't point out is how fast chicks grow, and how soon they need a bigger — and then an even bigger — cardboard box.

CHICKS HAPPEN

While I was brooding those chicks, I noticed that one of the buff bantam hens that came with the house had disappeared. Lo and behold, one day she reappeared, trailing a dozen fluffy yellow peepers. How did that happen? And hey — why can't I hatch chicks from my own hens instead of buying them at the farm store?

So began my quest to learn all about hatching the eggs of not only my chickens but also the ducks, geese, guineas, and turkeys that have populated my barnyard over the years. Pretty soon I became known as "the lady with chickens in her living room" — not the greatest idea because of all the dust baby birds make. One day I lucked out and ran across a used galvanized box brooder at a reasonable price, which allowed me to safely brood hatchlings in the garage. Later I found a deal on a couple of battery brooders, which let me

brood greater quantities of birds. I was in poultry heaven. My husband and I have since made a wide variety of brooders and have settled on custom designs that work best for us in our situation.

Meanwhile I learned how lucky I was to have a setting hen so early in the game, as not all modern hens are into motherhood. I have since looked after many a broody hen and her hatchlings and still occasionally get surprised by a stealth setter. Chapter 6 discusses the ins and outs of working with setting hens.

I have also learned to use a variety of incubators, starting with the minimalist galvanized steel models — affectionately known as tin hens — once sold by Sears, Roebuck, and working my way through an array of tabletop and cabinet incubators of many shapes, sizes, and features.

WHY HATCH AND NOT BUY?

So why would anyone bother hatching eggs in an incubator when hatchlings are so readily available from local breeders, farm stores, and mail order? Here are some of the many reasons:

- You wish to maintain a sustainable flock that acclimates to your particular locale
- You are working to restore an endangered breed
- Your chosen breed may not feature reliable setters or good mothers
- Your facility's conditions may not encourage broodiness
- Local wildlife or household pets may disturb the hens' nests
- You may wish to keep your hens laying, which stops when they start setting
- You are trying to produce the perfect show bird
- You want to raise hatchlings outside the normal brooding season
- You want more hatchlings than your hens are able to hatch
- You are seeking a fun and educational activity for your children or students

The decision to use a mechanical incubator is not necessarily either/or. I let my hens set whenever one gets the urge because I enjoy the sight of a hen with chicks. I use an incubator when I need a large number of chicks, such as to raise pullets as replacements for old layers and to replenish our family's broiler supply with the excess cockerels.

And sometimes combining brooding with artificial incubation makes sense, such as when a setting hen leaves the nest early, falls victim to a predator in midterm, or makes her nest in an unsafe environment for hatchlings. In such cases gathering up the started eggs and finishing the hatch in an incubator can save the day.

Between the fun of maintaining a breeder flock (or otherwise acquiring fertile eggs of the sort you want) and the joy of raising your own baby poultry comes a 3-to-4-week period of anticipation while you wait for the hatchlings to appear. Even old pros experience pangs of anxiety during this waiting period, because we know — mostly from past experience — all the things that can go wrong.

By paying careful attention to detail, starting with when and how you collect eggs destined for hatching, you can look forward to the enormously rewarding experience of hatching your own chicks. And should you wish to start your flock that way, skip right to chapter 6. While you're anxiously waiting for the eggs to hatch, you'll have plenty of time to prepare a brooder and read up on how to care for the adorable big-eyed fuzzballs that are about to enter your life.

Part 1
The Chicks

1

ACQUIRING YOUR FIRST CHICKS

Where you obtain your first baby poultry — whether they be chicks, keets, poults, ducklings, or goslings — will be determined to some extent by what breed or breeds you want and why you want them. Some sources are limited to one or only a few breeds; others offer a broad variety. And among all the potential sources, some carry strains that are more suitable for one purpose than another.

Selecting a Chicken Breed

The main purposes for keeping chickens are for eggs, for meat, for both eggs and meat, for exhibition, and for fun. Theoretically, these purposes are not mutually exclusive, but in reality a breed or hybrid that is considered suitable for both meat and egg production — called a dual-purpose or utility chicken — neither is as efficient at producing eggs as a layer breed nor grows as rapidly as a meat breed. Likewise, a strain that is developed primarily for exhibition generally does not lay as well or grow as fast as a strain within the same breed that has been developed for egg or meat production. Of course, if you want chickens just for fun, your choice is limited only by which breeds you most enjoy looking at.

You can learn more about the various available breeds and hybrids by visiting hatchery websites, by reviewing hatchery catalogs, and by reading a comprehensive book such as *Storey's Illustrated Guide to Poultry Breeds*. If your interest is primarily in exhibition, the American Poultry Association's illustrated and periodically updated *Standard of Perfection* is a must-have. Once you start homing in on a particular breed, seek out a regional or national club devoted to promoting that breed. Most such clubs have websites offering information and photographs on their chosen breed. Below is a quick review of the major breeds and hybrids, and the purpose to which each is best suited.

CHICKENS FOR EGGS

All hens, unless they are old or ill, lay eggs. The so-called laying breeds lay nearly an egg a day for long periods at a time. Other breeds lay fewer eggs per year, either because they take longer rest periods between bouts of laying or because they have a strong instinct to brood, and while a hen is incubating eggs she stops laying. Some breeds are nearly as prolific as those known as laying breeds, but they eat more feed per dozen eggs produced, and therefore are not economical as layers. The laying breeds share the following four characteristics.

They lay large numbers of eggs per year. The best layers average between 250 and 280 eggs per year, although individual birds may exceed 300.

They have small bodies. Compared to larger hens, small-bodied birds need less feed to maintain adequate muscle mass.

They begin laying at 17 to 21 weeks of age. Dual-purpose hens, by comparison, generally start laying at 24 to 26 weeks.

They do not typically get broody. Since a hen stops laying once she begins to nest, the best layers don't readily brood.

The most efficient laying breeds tend to be nervous or flighty. But kept in small numbers in uncrowded conditions, with care to avoid stress, these breeds can work fine in a backyard flock.

Ancona chicks

Chicken Breeds Typically Kept for Eggs

BREED	RATE OF LAY	SHELL COLOR	BREED	RATE OF LAY	SHELL COLOR
Ameraucana*	Good	Blue	Hamburg	Good	Tinted
Ancona	Best	Tinted	Lakenvelder	Good	Tinted
Andalusian	Better	White	Leghorn	Best	White
Araucana	Good	Blue	Marans*	Good	Brown
Australorp*	Good	Brown	Minorca*	Best	White
Barnevelder*	Good	Dark brown	Norwegian Jaerhon	Best	White
Campine	Good	Tinted	Penedesenca*	Good	Dark brown
Catalana*	Good	Tinted	Plymouth Rock*	Good	Brown
Chantecler*	Good	Brown	Rhode Island Red*	Good	Brown
Dominique*	Good	Brown	Rhode Island White*	Good	Brown
Empordanesa	Good	Dark brown	Welsumer*	Good	Dark brown
Fayoumi	Best	Tinted			

Good = 150–200; best = close to 300
*Better layers among the dual-purpose breeds

TYPICAL LAYER CHICKS

White Leghorn

Silver Lakenvelder

Blue Andalusian

Golden Campine

Silver Spangled Hamburg

Talking Fowl

breed. A genetically pure population having a common origin, similar body structure and other identifying characteristics, and the ability to reliably produce offspring with the same conformation and characteristics

brood. To sit on eggs until they hatch or raise a batch of hatchlings.

hybrid. A population parented by hens of one breed and cocks of another breed — generally for the purpose of increased efficiency in egg production or rapid growth for meat production — having similar body structure and characteristics but not able to always produce chicks with these body traits.

straightbred. Also called *purebred*; a chicken of a single breed that has not been crossbred with any other breed

strain. Also called a *line*; a related population of chickens — which may be either straightbred or hybrid — having nearly identical conformation and other identifying characteristics that make them especially suitable for a specific purpose, such as meat production, egg production, or exhibition

CHICKENS FOR MEAT

Any chicken may be raised for meat, but those best suited to the purpose were developed for rapid growth and heavy muscling. The meat, or broiler, breeds share four characteristics.

- **They grow rapidly.** These birds efficiently convert feed into muscle to produce economical meat.
- **They feather quickly.** Energy is spent building muscle instead of developing feathers.
- **Their bodies are deep and wide.** They have solid frames on which strong muscles grow.
- **They are broad breasted.** A sizable portion of white meat is harvested from these birds.

Breeds originally developed for meat include Brahma, Cochin, and Cornish. American dual-purpose breeds with the greatest potential for efficient meat production are the Delaware, New Hampshire, and Plymouth Rock. Although the Jersey Giant grows to be the largest of all chickens, it is not economical as a meat breed because it puts growth into bones before fleshing out, taking 6 months or more to yield a significant amount of meat for its size.

Our family raises chickens for both eggs and meat. Since we regularly hatch chicks to replace our older layers, and since approximately 50 percent of those chicks turn out to be males, we prefer a dual-purpose breed that both lays well and produces fairly hefty cockerels to fill our freezer.

Light Brahma chicks

Chicken Breeds Commonly Raised for Meat

AVERAGE AGE AT SLAUGHTER	BREED	AVERAGE WEIGHT COCKEREL		AVERAGE WEIGHT PULLET	
12–16 weeks	Delaware*	7½ lb	(3.4 kg)	5½ lb	(2.5 kg)
	New Hampshire*	7½ lb	(3.4 kg)	5½ lb	(2.5 kg)
	Plymouth Rock*	8 lb	(3.6 kg)	6 lb	(2.75 kg)
16–20 weeks	Ameraucana*	5½ lb	(2.5 kg)	4½ lb	(2 kg)
	Australorp*	7½ lb	(3.4 kg)	5½ lb	(2.5 kg)
	Buckeye*	8 lb	(3.6 kg)	5½ lb	(2.5 kg)
	Chantecler*	7½ lb	(3.4 kg)	5½ lb	(2.5 kg)
	Dominique*	6 lb	(2.75 kg)	4 lb	(1.75 kg)
	Faverolle*	7 lb	(3.2 kg)	5½ lb	(2.5 kg)
	Holland*	7½ lb	(3.4 kg)	5½ lb	(2.5 kg)
	Marans*	7½ lb	(3.4 kg)	6 lb	(2.75 kg)
	Rhode Island Red*	7½ lb	(3.4 kg)	5½ lb	(2.5 kg)
	Rhode Island White*	7½ lb	(3.4 kg)	5½ lb	(2.5 kg)
	Sussex*	7½ lb	(3.4 kg)	6 lb	(2.75 kg)
	Wyandotte*	7½ lb	(3.4 kg)	5½ lb	(2.5 kg)
20–24 weeks	Barnevelder*	6 lb	(2.75 kg)	5 lb	(2.25 kg)
	Crevecour	7 lb	(3.2 kg)	5½ lb	(2.5 kg)
	Dorking*	8 lb	(3.6 kg)	6 lb	(2.75 kg)
	Houdan*	7 lb	(3.2 kg)	5½ lb	(2.5 kg)
	Java*	8 lb	(3.6 kg)	6½ lb	(3 kg)
	Naked Neck*	7½ lb	(3.4 kg)	5½ lb	(2.5 kg)
	Orpington	8½ lb	(3.9 kg)	7 lb	(3.2 kg)
	Welsumer*	6 lb	(2.75 kg)	5 lb	(2.25 kg)
24 weeks (6 months) or more	Brahma	10 lb	(4.5 kg)	7½ lb	(3.4 kg)
	Cochin	9 lb	(4.1 kg)	6½ lb	(3 kg)
	Cornish	8½ lb	(3.9 kg)	6½ lb	(3 kg)
	Jersey Giant	11 lb	(5 kg)	8 lb	(3.6 kg)
	La Fleche*	7 lb	(3.2 kg)	5½ lb	(2.5 kg)
	Langshan*	8 lb	(3.6 kg)	6½ lb	(3 kg)
	Malay	7½ lb	(3.4 kg)	5½ lb	(2.5 kg)
	Orloff	6½ lb	(3 kg)	5 lb	(2.25 kg)
	Shamo*	9 lb	(4.1 kg)	5½ lb	(2.5 kg)

*Better growers among the dual-purpose breeds

Talking Fowl

chick. A baby chicken of either sex

cock. Also called a rooster; a mature male chicken

cockerel. A male chicken under one year of age

hatchling. A newly hatched bird of any species

hen. A mature female chicken

pullet. A female chicken under one year of age

DUAL-PURPOSE CHICKENS

Utility, or dual-purpose, straightbreds are the old-time farmstead chickens. They have four characteristics in common.

- **They produce a fair number of eggs.** Expect eggs-a-plenty, although the hens don't lay quite as efficiently as the breeds developed primarily for egg production.
- **Chicks hatched from their eggs will be pretty much like the parent flock.** They'll resemble their elders in temperament and body conformation.

- **Excess cockerels are worth eating.** However, they don't grow as efficiently as breeds developed primarily for meat production.
- **They are good to exceptional foragers.** They'll happily scrounge around in the yard or on pasture to obtain a great portion of needed proteins, vitamins, and minerals from bugs and plants.

TYPICAL DUAL-PURPOSE CHICKS

Barnvelder

Buff Orpington

Speckled Sussex

Dual-Purpose Chicken Breeds

Ameraucana	Holland	Penedesenca
Australorp	Houdan	Plymouth Rock
Barnevelder	Java	Redcap
Buckeye	La Flèche	Rhode Island Red
Catalana	Langshan	Rhode Island White
Chantecler	Marans	Shamo
Delaware	Minorca	Sussex
Dominique	Naked Neck	Vorwerk
Dorking	New Hampshire	Welsumer
Faverolle	Orpington	Wyandotte

HYBRID CHICKENS

Chickens developed for commercial meat production are hybrid — a cross between roosters of one breed and hens of another. Some of the best layers, particularly Leghorns, are not strictly hybrids but are a cross between roosters of one specialized strain and hens of another. Crossing breeds or strains produces chicks with hybrid vigor, which makes crossbred chicks stronger and healthier than either of their parents, resulting in better layers or faster growers than either breed or strain used in the cross.

Crosses are also made to take advantage of sex linkage, or characteristics that are genetically transmitted by genes carried on the sex chromosomes. These genetic characteristics allow chicks to be readily identified by gender from the moment they hatch by examining either color differences or rate of feather growth between the cockerels and the pullets.

Easter Egger

Among layers, the pullets may be easily sorted out and raised as layers, while the cockerels are discarded as being uneconomical to raise because, of course, they don't lay eggs and they put too little meat on the bone to raise for meat.

Among broilers, sorting the sexes results in groups of chicks that grow at a consistent rate — cockerels generally grow faster and are ready for the freezer sooner than pullets.

Fast-growing industrial Cornish-cross hybrids are the most efficient meat chickens for rapid growth and have white feathers for clean picking. Because their primary goal in life is to eat and grow, they do best in confinement. For pasture-raised broilers, hybrids are developed that have colored feathers, making them less visible to predators. These color range broilers are also a type of Cornish cross, but instead of being bred from industrially developed strains, they are bred from slower-growing traditional strains. These hybrids grow more quickly than their straightbred parents, making them more efficient for meat production, but they grow more slowly than industrial broilers, making their meat more flavorful.

A few commercial hybrids, particularly among the brown-egg layers, are considered to be dual purpose. However, if your reason for keeping a dual-purpose flock is self-sufficiency, which includes hatching eggs from your own chickens, hybrids are not the way to go, since they do not breed true, meaning they do not uniformly produce offspring that are exactly like the parents. The only way to obtain more of the same is to go back to the original cross.

Common Commercial Hybrid Chickens

LAYERS OF BROWN-SHELL EGGS	LAYERS OF WHITE-SHELL EGGS	COLOR RANGE BROILERS
		Black Broiler
Black Rock	Austra White	Cebe Black
Black Sex-Link*	California White	Cebe Red
Black Star	Leghorn strains	Color Yield
Cherry Egger		Colored Range
Cinnamon Queen	**LAYERS OF EGGS WITH VARIED SHELL COLOR**	Freedom Ranger
Golden Buff		Kosher King
Golden Comet	Easter Egger*	Redbro
Gold Star		Red Broilers
Indian River*	**WHITE BROILERS**	Red Meat Maker
Production Red	Cornish Cross	Redpac
Red Rock	Cornish Rock	Rosambro
Red Sex-Link		Silver Cross
Red Star		

*Dual-purpose strains suitable for both egg and meat production

TYPICAL HYBRID CHICKS

Left: Golden Buff, also called Cinnamon Queen, Golden Comet, Golden Sex-Link, Red Sex-Link, and Red Star. Right: Black Sex-Link, also called Black Star, Black Rock, and Red Rock

ORNAMENTAL CHICKENS

If efficient egg or meat production is not as important to you as having fancy chickens running around your yard, or you're hoping to get involved in exhibiting your chickens, consider the ornamental breeds. Although these breeds have been developed more for their aesthetic qualities than for the production of eggs or meat, some of them fall under the dual-purpose category. As a general rule, however, strains developed for exhibition purposes or as pets are bred more for appearance than for usefulness in producing eggs or meat.

TYPICAL ORNAMENTAL CHICKS

Gold Sebright

Black Frizzle Cochin

Sicilian Buttercup

Bantams

All bantams, because of their size and the small size of their eggs, are considered ornamental no matter how well they lay or how fast they grow. A bantam is a small chicken, about one-fourth to one-fifth as heavy as a large-size chicken and generally weighing 2 pounds (0.9 kg) or less. Not all bantams have the same ancestry as the large version of the same breed; some were developed from entirely different bloodlines to look similar, only smaller.

Nearly every large breed has a bantam version, but some bantams have no large counterpart. Bantams that do not are considered true bantams, while those that do are considered miniatures. But they are not exact miniatures — the size of their head, tail, wings, feathers, and eggs is larger than would be true of perfect miniatures.

Common Ornamental Chicken Breeds

		TRUE BANTAMS
Cochin	Phoenix	American Game Bantam
Crevecoeur	Polish	Belgian Bearded d'Anvers
Cubalaya	Redcap*	Belgian Bearded d'Uccle
Houdan*	Sicilian Buttercup	Booted Bantam
Kraienkoppe	Silkie	Dutch Bantam
La Flèche*	Spanish	Japanese Bantam
Langshan*	Spitzhauben	Nankin
Modern Game	Sultan	Rosecomb
Naked Neck*	Sumatra	Sebright
Old English Game	Yokohama	Serama
Orloff		

*Dual-purpose breeds

Care to Try Turkeys or Guineas?

Other popular species of land-dwelling poultry, or land fowl, include turkeys and guinea fowl. Both are, in many ways, similar to chickens. However, unlike chickens, which are generally fairly quiet — provided you don't keep a rooster — guineas and gobbling tom turkeys make enough racket to disturb neighbors. They are therefore better suited to rural areas than to populated neighborhoods.

Turkeys are basically meat birds, and all turkeys make good eating, although some breeds are considered to be primarily ornamental because they are small bodied, or grow slowly, or have dark feathers that do not readily pluck clean. Industrial strains, on the other hand, are developed for broad breasts, rapid growth, and light-colored pinfeathers. If you plan to hatch turkey eggs, be aware that the broad-breasted strains cannot mate naturally because of their enormous weight and outsized chests. Propagating a broad-breasted turkey requires artificial insemination.

Guinea fowl, too, come in large (French and Jumbo) strains and smaller common strains. French and Jumbo guineas are generally raised for the table; their dark, tender, lean meat tastes a bit like pheasant. Common guinea fowl are somewhat smaller, although they, too, make good eating when harvested young. Their more typical use is to reduce populations of ticks and other bugs, to ward off snakes, and to deter predators through mob attack. They are also interesting to look at, and their antics are fun to watch. By contrast to all the different breeds available among chickens and turkeys, the dozens of strains of guinea are distinguished by feather color, the most common color being pearl, consisting of medium gray feathers covered with a profusion of white spots.

Common Turkey Breeds

Farmstead Breeds

Bourbon Red

Bronze

Midget White

Narragansett

Broad-Breasted Breeds

Broad-Breasted Bronze

Broad-Breasted White

Giant White

Mammoth Bronze

Ornamental Breeds

Black (Spanish)

Chocolate

Jersey Buff

Royal Palm

Slate (Blue Slate)

White Holland

Keet   Poult

Talking Fowl

	MATURE FEMALE	MATURE MALE	BABY
Duck	Hen or duck	Drake	Duckling
Goose	Hen or goose	Gander	Gosling
Guinea	Hen	Cock	Keet
Turkey	Hen	Tom	Poult

Choosing Ducks and Geese

Waterfowl — ducks and geese — are similar to turkeys and guineas in that their noisy quacking and honking can annoy neighbors. They are therefore best suited for less populated areas, unless you opt for quackless ducks, otherwise known as Muscovies. Although waterfowl are among the easiest domestic poultry to raise, they should never be brooded together with chicks or other land fowl. Ducklings and goslings love to splash in water and can create quite a messy brooder, while chicks, keets, and poults need to stay dry to avoid getting chilled.

Like all domestic poultry, some waterfowl breeds, strains, and hybrids are kept primarily for egg or meat production, while others are largely ornamental. Among ducks the larger, meatier breeds do not lay as well as the midsize dual-purpose breeds or the smaller breeds kept primarily for eggs. Breeds with white or light-colored feathers pluck cleaner, although the breeds with more colorful plumage are less visible to predators.

Among geese no breed is as prolific as the best layers among ducks. Most domestic goose breeds have been developed for meat, although some are primarily ornamental because of their small size. Mature geese are strong birds that can become angry when teased and aggressive during nesting season. The larger strains of Toulouse sometimes have trouble breeding naturally.

TYPICAL DUCKLINGS AND GOSLINGS

Fawn & White Runner   White Crested Duck

Pilgrim Geese

Wild Flying Mallard   Toulouse Goose   American Buff Goose

Waterfowl Breeds

DUCKS	GEESE
Lightweight	**Lightweight**
Campbell	Chinese
Harlequin	Roman
Magpie	Shetland
Runner	
Midweight	**Midweight**
Ancona	American Buff
Buff (Orpington)	American Tufted Buff
Cayuga	Pilgrim
Crested	Pomeranian
Swedish	Sebastopol
Heavyweight	**Heavyweight**
Appleyard	African
Aylesbury	Embden
Muscovy	Toulouse
Pekin	Toulouse, Dewlap
Rouen	Toulouse, Giant
Saxony	
Bantam	**Ornamental**
Australian Spotted	Canada
Call	Egyptian
East Indie	
Mallard	

What Is a Heritage Breed?

Heritage, or heirloom, breeds are traditional breeds developed over many years of selective breeding. Such ancient breeds as Sweden's downy Hedemora chicken, Japan's longtail Onagadori chicken, and Scotland's hardy Shetland goose certainly qualify. However, the word "heritage" has no precise definition that everyone fully agrees on.

The American Livestock Breeds Conservancy has developed a list of criteria for use in labeling eggs or meat sold as being from a heritage breed. According to this definition, a heritage breed must:

- Have been recognized by the American Poultry Association prior to the mid-twentieth century
- Reproduce through natural mating
- Have the genetic ability to live a long, vigorous life
- Thrive outdoors under pasture-based management
- Have a moderate to slow rate of growth for proper development

Finding the Hatchlings for You

Once you decide what kind of hatchlings you want, it's time to seek out sources. Some sources are highly specialized; others offer a broad variety of species and breeds. Some sources produce the same breeds year after year, while others introduce one or more exotic new breeds each year.

If you're getting poultry mainly for the fun of it, you needn't be as fussy about finding a source as you would be if you want birds that will produce eggs or meat economically or if you wish to be successful at showing. Below are the primary sources for chicks.

FARM STORE LIMITS

Not every farm store handles chicks, and those that do offer a limited selection for a limited time; chicks are generally available there only from early March through Easter week. Another disadvantage to buying at a farm store, especially at some big chains, is that the employees don't know, or are not allowed to tell you, what kind of chicks they are offering. An advantage is that you can get as few as you want or as many as they have.

One thing you want to avoid is getting free chicks when you buy a sack of chick starter ration.

Those chicks are commonly Leghorn cockerels that most people have no use for, as they don't put enough meat on their bones to make good eating — and who needs a coop full of roosters?

Some farm stores, particularly the chains, obtain their chicks from hatcheries certified by the National Poultry Improvement Plan (NPIP; see box). Other farm stores, especially ones that are locally owned, acquire their chicks from local farmers who may or may not participate in NPIP.

National Poultry Improvement Plan

The National Poultry Improvement Plan (NPIP) is a federal program that works through state programs to certify participating poultry producers as having flocks that are free of several serious diseases. Birds purchased from an NPIP participant are reasonably certain to be healthy. On the other hand, a lot of poultry breeders don't want to get involved in government bureaucracy, which does not automatically mean their chickens are unhealthy. Use your best judgment when dealing with a breeder that does not participate in NPIP. To search the NPIP directory of participants, visit www.aphis.usda.gov/animal_health.

Partridge Rock

LOCAL POULTRY KEEPERS: A GOOD START

Lots of people get their first chicks from a neighbor, which has the advantage that the strain is already acclimated to your local environment compared to birds you might purchase from some distant place. Connecting with local poultry keepers gives you someone to contact when you have questions or concerns about raising your own birds. Buying locally also gives you a chance to see the parent flock and verify for yourself that they are maintained in a sanitary and healthful condition.

Your local farm store may be able to tell you who has chicks for sale or may have a bulletin board where customers can post livestock for sale. Your county Extension agent should know who keeps poultry in your area and may know a 4-H member with chicks for sale. The county or local fair poultry show, if your county has one, is another place to meet local poultry keepers and is a good place to find out if your area has a regional poultry club whose members can be invaluable in helping you get started. Also check local free shopper newspapers, radio swap meets, and local networking websites, including the Craigslist classified ads (www.craigslist.org) for your locale. At those same places you might also post a wanted notice describing what breed you are looking for and how many hatchlings you are seeking.

KNOW YOUR LOCAL HATCHERY

A hatchery is a place that incubates eggs and sells the resulting chicks. In some cases the hatchery doesn't own the breeder flocks that lay the eggs but contracts with local farmers to obtain eggs from the desired breeds. In other cases the hatchery may own the breeder flocks and contract with local farmers to care for them, giving the hatchery somewhat greater control over the quality of the breeder flocks.

Not many of us live near enough to a hatchery to stop in to pick up a batch of chicks, and some hatcheries don't allow visitors, to prevent the introduction of poultry diseases into their facility. Those hatcheries that allow visitors are not likely to have birds on display and certainly are not likely to allow visitors into the hatching room. Even though a hatchery is nearby, call ahead to determine if they have what you want. You may have to order your chicks well in advance of when you want them, as hatcheries typically incubate eggs to fill orders rather than hatch eggs and then try to sell the chicks.

Some hatcheries are highly specialized, for example, at producing Leghorn pullets for local egg farms. Other hatcheries may deal solely with waterfowl, or turkeys, or guinea fowl. Unfortunately, a few hatcheries churn out large numbers of low-quality birds. Most hatcheries, however, do participate in NPIP.

Light brown Leghorn

HATCHERY CHICKS FOR SALE

Hatcheries offer several options for you to choose from. Among them are whether you want your chicks to be sexed or unsexed, vaccinated, or debeaked.

Sexed chicks have been sorted according to whether they are male or female, so you can get as many pullets or cockerels as you want, with as much accuracy as current technology allows. Except among broiler strains, within a given breed sexed pullets cost the most, straight run cost less, and sexed cockerels cost the least. In most cases cockerels have the least value because a flock needs fewer roosters than hens, or none at all, if you or your neighbors don't want to hear crowing. You do not need a rooster to get eggs, and roosters can be rough on the hens, although many people feel that including at least one cock in the flock doubles the enjoyment of having chickens. Among commercial broiler strains, cockerels are the most expensive and pullets are less expensive than straight run, because the cockerels grow better than the pullets.

Unsexed chicks — also called *as-hatched* or *straight run* — have not been sorted by gender and therefore are mixed exactly as they hatch. Theoretically, a hatch should be 50/50. Some hatches have more pullets than cockerels, but more often the ratio is 60/40 or even 70/30 in favor of cockerels. The smaller the number of chicks, the more likely the ratio will be skewed one way or the other, most commonly in favor of cockerels. The greater the number of chicks, the better chance you have of getting at least 50 percent pullets. If you plan to hatch eggs in the future, getting a batch of straight-run chicks will give you hens to raise for eggs, plenty of roosters to choose from to fertilize the eggs, and surplus cockerels to put in the freezer.

Vaccinations offered by some hatcheries include immunization against such things as Marek's disease or coccidiosis. Marek's is a fairly common disease caused by a highly contagious herpes virus that results in tumors, paralysis, and suppression of the immune system; it does not affect humans. Some chicken breeds and strains are genetically resistant to this disease in varying degrees. The vaccine helps reduce losses but does not confer complete immunity. In any case, vaccinating broilers is unnecessary because they are harvested before they reach the age when Marek's becomes an issue.

Coccidiosis is a common protozoal disease that affects chicks but not humans. Vaccination stimulates a natural immunity that produces lifetime protection against this disease, but you must take care to never feed your chicks medicated rations. They contain a coccidiostat, which is a drug that inhibits the development of coccidiosis and therefore would neutralize the vaccine. An alternative to vaccination is to manage your chicks in a way that helps them avoid getting coccidiosis.

Debeaking, or surgically shortening the beak, is another option offered by some hatcheries as an industry practice designed to prevent overcrowded chickens from eating one another. Debeaking mars a bird's appearance. Furthermore, with proper management cannibalism should not be a problem, making debeaking unnecessary. You can find more information on Marek's disease, coccidiosis, and cannibalism in chapter 5.

MAIL-ORDER CHICKS

Mail-order sources for chicks abound, and nearly every one has a website or a catalog, or both. Trying to pick a reputable mail-order source out of the crowd can be daunting. Some of them are hatcheries; others are brokers that take orders and outsource fulfillment. I once unwittingly bought some goslings from such a firm, and a year or two later when I wanted more of the same, the folks at this firm said they couldn't remember, and had no record of, who had shipped the first goslings.

One way to find a reputable source is to chat with the manager of a local farm store that sells chicks each spring. By doing that I obtained a short list of reliable mail-order sources, as well as advice on some sources to avoid. Personally, I like a source that publishes photographs of real chickens, rather than idealized artwork.

Plan to place your order around the first of the year, especially if you want a rare or popular breed. Some breeds sell out rapidly; if you delay ordering, you run the risk that your chosen source will sell out, and you will have to either find an alternative source or select a different breed.

A few mail-order firms specialize in shipping small batches, but most require a minimum order, usually 15 or 25, to ensure that enough birds are packed into the shipping box to keep each other warm during the long journey. A large box holds about 100 chicks, or the equivalent in other species; a medium box holds 50; and a small box holds up to 25. Fewer than 15 chicks are not enough to stay warm, but some firms successfully ship as few as 3 along with a heating pad.

If you live in a remote area, check with your local postmaster to make sure chicks may be shipped there. Chicks can do okay for 2 to 3 days without food and water, while they're still surviving on yolk reserves. If the journey from the hatchery to your destination would take longer than 72 hours, the postal service will not accept the shipment.

Assuming your post office handles live birds, notify the personnel there that you have ordered chicks, tell them when you expect the chicks to arrive, and ask that you be called when they get there; most hatcheries will post your phone number on the outside of the box. You can expect your chicks to arrive within 1 or 2 days of being shipped, which is usually on a Monday, so chicks won't have to endure being left in a closed post office over a weekend. Be prepared to pick up your chicks at the post office so they won't have the extra stress of riding around in the carrier's vehicle and so you can quickly get them home to feed, water, and warmth.

When you receive chicks by mail, open the box in front of the mail carrier to verify any claim you may have for losses. Occasionally, chicks die in transit, either because they weren't vigorous to start with, or because they were mishandled along the way — perhaps handled roughly; left for too long on an extremely hot or cold dock; or misrouted, thus taking longer than usual to be delivered. But most of the time chicks arrive in good condition, cheeping loudly because they are tired, hungry, thirsty, and eager to get to their new home.

BARGAIN ASSORTMENTS

Some hatcheries offer assortment packages that can be a real bargain if you're not fussy about what you receive. Because not all eggs hatch, a hatchery typically incubates extra eggs to ensure it can fill all orders. Assortment specials made up of any extras that hatch might be packaged as large-breed assortments, bantam assortments, waterfowl assortments, surprise (anything goes) assortments, or whatever assortment the hatchery feels will move surplus hatchlings.

Since most people want to start chicks in the springtime, typically around Easter, some hatcheries offer bargain assortments as the hatching season winds down. Before placing your order, read the fine print to make sure the chicks are what you really want. An assortment of large-bodied cockerels won't be much of a bargain if what you want is a coop full of small-bodied laying hens.

Chicks are shipped in boxes with lots of holes for ventilation. A midsize box like this one holds about 50 chicks or the equivalent in other species.

SPECIALTY BREEDERS

Most small-scale poultry breeders specialize in one breed, or sometimes several similar or related breeds. A specialty breeder would be the way to go if you are seeking a highly specific type of poultry. For instance, if you plan to exhibit your chosen breed, you will want to seek out someone who raises and successfully shows that breed. Not all exhibitors are willing to part with chicks, but most exhibitors know others who specialize in the same breed and may have what you are looking for.

Most breeds have one or more specialty clubs whose members raise, show, and promote their chosen breed, and many such clubs publish a membership list to help you locate members in your area. You can find out if any clubs promote your chosen breed by doing an Internet search using as your key words the name of the breed followed by the word "club"; if nothing comes up, substitute the word "association" for "club."

If you wish to pasture poultry, look for a source that specializes in producing layer strains or color range broilers suitable for pasturing. Some of the bigger hatcheries offer a limited selection of such breeds. Other sources may be found through any of the organizations that promote poultry pasturing. And while you're shopping around, keep in mind that breeders that handle poultry on a commercial level, including for food production, are more likely than breeders who raise poultry for show to participate in NPIP.

LIVESTOCK AUCTIONS AND SWAP MEETS

Local livestock auctions and swap meets are good places to check current prices, meet breeders, and learn who has quality stock and who doesn't. But they are also the best places to avoid buying poultry, because birds (and their diseases) are brought together from multiple sources. Hatchlings may look perfectly healthy on site, but by the time you get them home, they may not do so well. It's not worth the heartbreak or the possibility of introducing diseases onto your place that might affect healthy chickens you may acquire later on. Furthermore, even if you find chicks for sale, you might not be correctly told what breed they are.

Swap meets are a bit better than livestock auctions, because you are more likely to be told the correct breed and the chicks won't be handled as roughly, but the same caveat applies regarding the spread of diseases. I don't care to attend auctions or swap meets in case I might bring home some disease pathogen in the form of poop on the bottom of my shoes. The same is true of buying used equipment at these places.

The potential problems created by sales that bring together birds from many sources may be avoided by using one of the online auctions, which may be found by doing a keyword search for "online poultry auctions." Most of the poultry forums also have a section where chicks may be offered for sale. Unless the seller is well known in the poultry community, though, you may not get what you think you're buying, and you have no way of knowing if the seller runs a clean operation.

HATCH YOUR OWN

Getting started with poultry by acquiring fertile eggs and hatching them yourself is a big undertaking, and therefore one I don't heartily recommend. However, the idea does have a certain logic, not to mention charm. It definitely makes a terrifically educational project for a classroom or a home-schooling family. If you're determined to go that route, you can find information on acquiring eggs for hatching and choosing an incubator to hatch them in, by jumping to part 2, which starts on page 109.

The Right Time for Chicks

Hatchlings of one sort or another are available nearly year-round, although most sources hatch primarily in the springtime for these reasons:

- Some breeds lay eggs only in the spring
- Demand for hatchlings is highest in the spring
- Breeder flocks produce the healthiest chicks in spring
- Chicks thrive best in warming spring weather

Although the greatest selection in hatchlings is available from about February through June, March and April are ideal months to brood chicks because the weather is starting to warm up but is still cool enough to discourage diseases. As the weather gradually warms, you can reduce the amount of necessary artificial heat accordingly, which helps chicks feather out more quickly, so they'll be fully feathered against cold weather come the following winter. Furthermore, spring pullets will start laying in the fall and continue laying throughout the winter. Their production will peak the following spring and from there will gradually decline as the year progresses.

If you're raising commercial-strain broilers, you'll need to avoid the stressful heat of summer. Since they take only 6 to 8 weeks to reach harvest weight, either start them early enough in spring to have them in the freezer before hot weather hits or start them in late summer so they will grow out during cooler fall weather. Raising one batch in spring and another in fall spreads out the workload of managing and harvesting broilers for the family freezer. If, for example, your family eats one chicken per week, or about 50 chickens per year, you might raise two dozen in the spring and another two dozen in the fall, which not only spreads the workload over time but also requires less freezer space for storage.

Signs of spring

2

SETTING UP YOUR BROODER

Before the arrival of hatchlings, have your brooding facility set up and ready for them to move into. Setting up a brooder isn't difficult, or expensive, once you know what's needed. Although fully appointed brooders are available for sale, a sturdy cardboard box serves quite nicely as a first-time brooder.

Brooding baby poultry of any species generates a lot of fine dust, so locate your brooder where that won't be a problem. Your living room or guest room, for example, is not ideal because the dust will settle on drapes, furniture, and carpets. Much more suitable is a room with easy-to-clean hard surfaces, such as a laundry room or spare bathroom. A garage or tool shed is another possibility, provided it isn't drafty. If you have an unoccupied coop or other outbuilding where the birds will live when they mature, you might set up the brooding facility there.

Brooder Features

Hatchlings are not entirely helpless, but until they grow a full complement of feathers you'll need to keep them warm and dry and protect them from harm. Like any other babies they also must be fed and kept clean. A brooder serves all the necessary functions by means of the following features:

Adequate space for the number of birds. Initially, they don't need much room, because (like other babies) they spend most of their time eating and sleeping. If you have a larger brooder you can block off a portion of it for the first days to keep the baby birds close to the sources of food, water, and heat. As the birds grow and become more active, they will need more room, both to avoid stress and conflicts and to prevent a too-rapid buildup of moisture and manure. Provide additional space either by moving them to larger quarters or by opening up the portion of the brooder you blocked off.

A reliable and adjustable heat source. A hatchling's body has little by way of temperature control, although a group of them can stay warm by huddling together in a small space — which is why a boxful of babies may be shipped by mail. Given sufficient space to move around within a brooder, chicks need an external source of warmth until the down covering their bodies gives way to feathers, starting at about three weeks of age. As they grow they need less and less external heat, because their bodies gradually generate more warmth that helps heat up the brooder. The brooding temperature must be systematically reduced as they grow.

Escape-proof design. Hatchlings can sneak out through the smallest openings at the brooder's floor level. A brooder that lacks a sliding adjustable gate to prevent escape should be solid for at least 6 inches (15 cm) up from the floor. At about three weeks of age, the chicks will start exercising their wings and try to perch on anything their tiny wings let them fly onto. Chicks kept in a brooder with low walls and no cover may inadvertently escape when they perch on the brooder's edge and then come back down on the wrong side.

Protection. The safest sort of brooder is one that entirely encloses the chicks and must be opened purposely to tend to their needs. Inside the family home, exuberant children and excited pets can harm delicate chicks without meaning to. A brooder set up in a garage or outbuilding must offer protection from such predators as weasels, rats, and snakes that can squeeze through incredibly small openings.

Freedom from drafts. A brooder inside the house most likely won't be subject to drafts, provided it is not directly in the air current generated by a fan or air conditioner. A brooder in a garage or outbuilding must be shielded from any drafts that might reduce the brooding temperature and chill the birds.

Good ventilation. A brooder needs good air exchange to provide plenty of oxygen while removing the carbon dioxide and moisture that chicks generate through breathing. For this reason

Talking Fowl

brooder. A facility designed to replace a mother hen as a place where hatchlings are temporarily raised until they have enough feathers to keep themselves warm

droppings. Bird poop

forage. To wander around a yard or on pasture looking for food; or the food itself

grow unit. Also called *grow house, grow pen,* and *halfway house.* Intermediate housing for birds that have outgrown their brooder but are still too small to be safe and comfortable in facilities designed for full-grown birds

hatchling. A bird that has recently hatched from an egg

picking. Undesirable land-fowl behavior involving pulling out each other's feathers or pecking at flesh to the point of causing open wounds

most brooders have either a screened top or a sliding adjustable gate around the sides to provide adequate ventilation.

Protection from moisture. Chicks not only generate moisture by breathing but also by pooping and by splashing the drinking water. The moisture issue is worse for waterfowl. Brooder bedding that accumulates moisture becomes moldy, creating a health hazard for hatchlings, and therefore must be cleaned out and replaced often.

Proper flooring for baby birds. The brooder floor should not be so slippery chicks can't walk without sliding, nor should it be covered with a type of bedding chicks will trip over, have trouble walking on, or be attracted to eat. The suitability of bedding and flooring changes as chicks grow. This subject is covered in detail in chapter 3.

Sufficient light to find the feeder and drinker. If the source of heat is a lightbulb or heat lamp, no additional light source is needed. If the heater does not also provide light, you must

light the brooder in some way, such as light in the bright room in which the chicks are brooded, sunlight coming through a window, or a low-level lightbulb in or near the brooder.

Suitable feeders and drinkers. Different styles of feeders and drinkers are available that have been designed specifically for baby poultry. This subject is covered in detail in chapter 3.

Keep Spare Bulbs

Brooder heaters consisting of light-emitting bulbs may burn out. Some ready-made brooders use an ether-filled wafer to regulate heat, but the wafer may spring a leak (see page 134 for information on wafers). Depending on the type of heat your brooder uses, keep one or more spare bulbs or wafers on hand to ensure your baby birds don't get chilled should the first one fail.

Solid Floor versus Hardware Cloth

Some brooders have a solid floor; others have a floor made of hardware cloth. A solid floor provides a surface to hold a layer of bedding, which helps keep chicks dry and insulates the brooder floor for improved warmth. A solid floor also gives chicks an opportunity to develop natural immunity through gradual exposure to common microbes in the environment, particularly coccidiosis.

Hardware cloth is easier to clean, since droppings and other debris fall through the mesh to be collected on a droppings tray or a layer of newspaper for easy disposal. Small-mesh hardware cloth is ideal for waterfowl because it allows splashed drinking water to drain away from the birds. But for land fowl, hardware cloth has these disadvantages:

A resting bird can get a hock joint jammed into the mesh. If the bird is not immediately extracted, the joint will twist or swell and the bird may become permanently crippled.

Wire mesh is hard on a bird's feet. As chicks — especially the heavier breeds — grow

and gain weight, they may develop sores on the bottoms of their feet.

Hardware cloth does not give chicks gradual exposure to coccidiosis, which helps them develop immunity. Chicks that are later moved to a solid floor may therefore suffer an outbreak of coccidiosis.

Chicks brooded on wire are at greater risk for cannibalism than chicks raised on litter. Since exploratory pecking is normal behavior, chicks with nothing else to peck will peck each other, in which case pecking becomes picking.

Hardware cloth may be used — briefly and cautiously — in grow housing for birds that no longer need a heated brooder but are still small and lightweight enough that they don't get foot sores. A plywood or other hard surface, for example, might be fitted over a hardware cloth floor for the first few weeks of brooding, then removed when the chicks are big enough to walk on wire mesh but still small enough to remain in confinement.

Ready-Made Brooders

Ready-made brooders come in many different styles and sizes, ranging from small ones that hold a dozen or so chicks to large outfits designed to brood hundreds. The style and size that are best for you will depend on how often you plan to brood, how many birds you plan to brood at once, and how much money you want to spend.

BOX BROODER

A commercially available box brooder is made of metal or metal and plastic. It typically has a built-in heat source, a low-level light, and a hardware-cloth floor with a removable tray underneath to collect waste. Lining the tray with sheets of paper makes the trays easier to clean, and the paper may be composted. Built-in feed and water troughs are attached to the outside along three sides to furnish plenty of feeding and drinking space, while keeping the chicks from fouling the feed or water.

A sliding adjustable gate protects each trough so chicks can't climb into and drown in the water, poop in the feed, or fall out of the brooder. To ensure the chicks can always reach feed and water, the bars must be monitored regularly and adjusted as the chicks grow.

Typical box brooders are sized to hold about 50 chicks for 4 weeks or 100 chicks for 2 weeks. Before the chicks grow so tall that their heads rub against the top, or so big that they can't get their heads through the bars to eat and drink, they must be moved to a grow unit.

GQF Manufacturing's aluminum universal box brooder will brood 100 chicks for 2 weeks or 50 chicks for 4 weeks.

A box brooder has troughs around the outside to hold feed and water and protective bars to ensure chicks can eat and drink but not escape.

BATTERY BROODER

A battery brooder consists of a series of box brooders stacked on top of each other. The removable droppings trays prevent waste generated at one level from falling on chicks in the level below. A battery lets you brood lots of chicks in a small space, and the various levels offer a convenient way to separate birds of different genders, ages, breeds, or species.

Some batteries are designed with a rack that holds a specific number of levels, typically three to five. Others consist of a wheeled frame onto which you can stack individual box brooders as high as you can reach. Each level has its own heater, so you can save electricity by turning off heat in any unoccupied level.

Because a box brooder can accommodate chicks for no more than about 4 weeks, some batteries consist of a combination of box brooders and grow units. Each grow unit is similar to a box brooder but lacks a heater and is somewhat taller to accommodate growing birds.

A typical scenario for a battery consisting of one brooder and two grow units (or, for raising broilers, four grow units) is to brood four dozen chicks for 4 weeks, after which they are split up and moved to the grow units. Another four dozen chicks then may be started in the brooder. By the time the second batch is ready to be moved to the grow units, the first chicks are big enough to move outside or, if they are broilers, have reached the age of harvest.

A battery filled to capacity tends to get· messy fast, and batteries are difficult to clean. Furthermore, they work best when located in a heated room or an outbuilding that is free of drafts.

This battery from GQF Manufacturing has a brooder at the top and two grow units below.

AREA BROODER

The most common type of brooder used on poultry farms is an area brooder, which is not an individual brooding unit but a small space separated from a larger area for the purpose of confining chicks close to heat, feed, and water. Area brooders have traditionally been used to brood large numbers of chicks on a farm, but the same concept easily may be scaled down to brood just a few chicks.

An area brooder has an advantage over a box or battery in being easily expandable, so you can brood more chicks at a time and they won't outgrow their housing as quickly, if at all; the building in which the area brooder is set up may be the same building in which the birds will live when they mature. An area brooder gives chicks flexibility to adjust their own comfort level, especially when days are hot and nights are cool.

Kits are available that include all the necessary parts to set up a small area brooder. For more than just a few chicks, an area brooder is generally purchased as separate parts that are arranged according to the number of chicks to be brooded. Area brooders are therefore described more fully under Homemade Brooders on page 36.

An area brooder starter kit includes a feeder, a drinker, a cardboard corral, a heat bulb and fixture, and a thermometer — just add chicks.

Announce Your Approach

Most ready-built brooders are designed for approach from the side, while most homemade brooders are approached from the top. Baby birds may panic if they don't see you coming until you reach into the brooder to replenish their feed and water. After all, most predators descend from above to snatch up a little chick. To avoid stressing out your baby birds when approaching them from the top, pause a moment to hum, sing, talk, or whistle to alert them before you reach in.

Homemade Brooders

A brooder need not be an expensive, commercially built affair or even be a permanent fixture. As long as you maintain the principles of security and warmth, the possibilities for brooding chicks are limited only by your imagination.

A SIMPLE CARDBOARD BOX

Many a chick has been brooded in a sturdy cardboard box, which has the distinct advantage that it may be disposed of and replaced, instead of having to be cleaned and stored. Roomy boxes sometimes may be obtained from a grocery store and may be purchased from a moving company or an office supply store.

Turn the top flaps into the box to make the sides rigid and keep them from sagging under the weight of whatever you use for a cover. Adding a cover is always a good idea to keep chicks in and other critters out. For good ventilation the cover should not be a solid object. Ideal covers include a piece of hardware cloth or chicken wire, short lengths of vinyl-coated wire shelving, or an old refrigerator or oven grill. Such a cover may be attached easily to the box with hinges made from string or wire threaded through holes poked around the top of the box. The heat source may be securely attached to the top of the cover or hung from the cover down into the box, depending on the size of the box and the degree of heat needed.

Since chicks generate a lot of moisture, keep the bottom of the box from getting soggy by lining it with additional cardboard, multiple layers of newspaper, or a large plastic trash bag before adding any bedding. Add a heater, a feeder, and a drinker, and you're ready for chicks.

A cardboard box is not the greatest option for brooding waterfowl, because little ducklings and goslings like to play in their drinking water and can rapidly turn the cardboard to mush. With a little ingenuity, however, you can adapt a cardboard box for use by waterfowl. For instance, you can outfit the box with a drain underneath the drinker to channel excess water into a 5-gallon (19 L) bucket. The box, of course, will be trashed beyond reuse, but the drain assembly may be used from year to year.

A cardboard box makes a serviceable brooder that may be disposed of when no longer needed.

Reused Housing

If you put chicks or growing birds into a brooder or any other housing that has held poultry in the past, the facility must be thoroughly cleaned, swept free of dust and cobwebs, and washed down. First clean it with warm water and detergent, then with warm water and chlorine bleach (¼ cup bleach per gallon of hot water; 15 mL/L) or other disinfectant approved for use with poultry. Leave the facility open to air dry completely before bringing in the new birds. Ideally, clean the unit as soon as you have removed the previous birds, not leaving it caked with dirt until the new residents are ready to move in.

WHATEVER WORKS

garden cart

baby crib

satellite dish

bureau or cabinet

bathtub

Repurposed Items: Be Creative!

Do an Internet keyword search for chick brooders and you will find all sorts of imaginative possibilities. Anything that satisfies all the requirements of a successful brooder should work fine. Chicks have been brooded in such things as a garden cart, a baby crib, a satellite dish, a large fish tank, and a redesigned cabinet or bureau.

The spare-room bathtub makes a reasonable choice for waterfowl, as they can easily be allowed the treat of an occasional shallow swim under your supervision. Lots of people brood chicks in a kiddie wading pool lined with bedding, although you can't easily fit it with a predator-proof cover. Also, chicks will perch on the edge and escape unless a chick corral is added to increase the height of the sides.

TRY A PLASTIC TOTE

An extra-large plastic storage tote, in the 100-gallon range, makes an easy-to-clean brooder, and the snap-on lid secures it from predators. The advantage over a cardboard box is that the tote won't be demolished by moisture. Moisture does tend to collect in the bedding, however, because it has no place else to go, so the bedding must be cleaned out and replaced every other day or so, to keep it from getting moldy. For brooding ducklings or goslings, an elevated floor made of hardware cloth (with all the cut edges turned downward so the birds can't snag a foot) gives excess moisture someplace to go until you come along to mop it out. For all species, having two totes is handy, so you can move the little ones to a clean, dry tote while you clean and dry the first one.

A plastic tote can easily get too hot, so keep a close eye on the birds' comfort level and adjust the heat accordingly. For air circulation, and to prevent overheating, a ventilation hole must be cut into the tote's lid, which is not an easy job but may be neatly done with a utility knife and plenty of elbow grease. Hardware cloth fastened to the lid will keep out cats and other predators. The heater may be hung from the hardware cloth inside the tote, or placed on top of the hardware cloth outside the tote, depending on how much heat it generates and how much heat is needed.

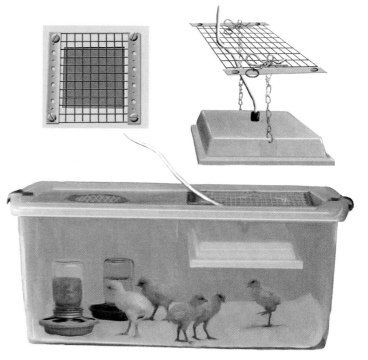

An extra-large storage tote, well managed to prevent a buildup of moisture and excess heat, holds about two dozen chicks for 1 week or up to a dozen for 2 weeks.

STOCK TANK HOMES

A popular brooding option is a livestock watering tank, also called a watering trough, which may be hosed out between uses. Stock tanks, purchased at a farm store, are either round or oval, come in various sizes, and are either plastic, rubber, or galvanized steel. If frequently used for brooding, a tank of plastic or rubber works better than a galvanized tank, because eventually the bottom of a galvanized tank will rust out from moisture trapped in the brooder bedding. A tank with a built-in drain is especially handy for brooding waterfowl.

A piece of hardware cloth or chicken wire secured on top provides ventilation while keeping out cats and other bird eaters and prevents growing chicks from flying out. If necessary to eliminate drafts, lay an empty feed sack or a piece of cardboard or plywood across part of the top.

A plastic or rubber stock tank makes a sturdy and roomy brooder that won't rust from chick-generated moisture.

WHAT'S TO KNOW ABOUT WIRE MESH?

HARDWARE CLOTH

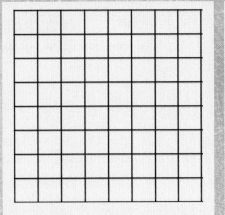

hardware cloth (⅛ inch [3 mm])

hardware cloth (¼ inch [6.5 mm])

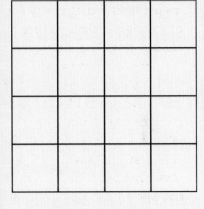

hardware cloth (½ inch [13 mm])

HEXAGONAL NETTING

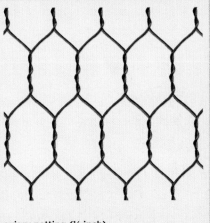

aviary netting (½ inch)

chicken wire (1 inch [25 mm])

Hardware cloth is firm and rigid, and works well for brooder floors, drinker platforms, incubator hatching trays, and pedigree baskets.

Hexagonal netting is cheaper and more flexible and is easy to fashion into brooder covers.

Either kind of wire may be used for brooder sidewall ventilation openings, although most hatchlings can walk right through chicken wire if it is applied too close to the bottom of the brooder.

WOODEN BROODER

By constructing a brooder out of wood, you can make it any size that fits your needs, for use either inside or out, either portable or stationary, and flexible enough to use with land fowl or waterfowl. A common design is basically a wooden box on legs with a hardware cloth floor. A piece of cardboard or plywood laid over half the floor for the first week or so will give hatchlings a comfortable place to rest. A window screened with hardware cloth at the front will let the birds see anyone approaching the brooder, making them less likely to be startled and panic.

Newspapers or opened-out feed sacks spread beneath the box collect droppings that fall through the hardware cloth. Alternatively, a shallow under-bed storage tote may be used as a droppings pan, which has advantages for brooding ducklings and goslings, as it holds the water they splash. A removable floor, with the brooder walls fitted tight against the tote, allows you to brood chicks and other land fowl on bedding rather than on wire mesh, and the plastic tote is easy to clean for reuse.

AREA BROODERS GROW WITH CHICKS

If you have a predator-safe and draft-free place to set it up, creating an area brooder is one of the simplest and least expensive brooding options. Because it is expandable, it eliminates the need for an intermediate grow pen. In addition to feeders and drinkers, an area brooder consists of two essential parts — a canopy heater and a brooder guard.

Canopy Hover Heater

The heater is either hung from the ceiling or raised up on short legs, and the canopy directs warmth close to the floor. Because the canopy hovers over the chicks like a mother hen, it is traditionally called a hover. If the area brooder is set up in a really cold place, curtains may be hung around the edges of the hover to keep in heat and keep out drafts — although curtains tend to inhibit the chicks' movement.

A plywood or cardboard hover may be fitted with adjustable legs. If the box is cardboard, staple a second set of 1×2s to the outside corners to hold the bolts.

The concept of a hover is to provide a warm area where chicks can rest and sleep. Ready-built hovers are available for brooding sizable numbers of chicks. For just a few chicks, a low-hanging heat lamp or heater panel works fine. For a larger number of chicks, or for chicks brooded in cold weather, the heat may be concentrated under a hover fashioned from a cardboard box or constructed of plywood. The box must be of sufficient size to avoid close contact with the heater, to prevent a fire. A 2-foot-by-2-foot (0.6 × 0.6 m)-square hover is sufficiently large to accommodate 12 to 15 chicks for the entire brooding period.

A heat source (as described on page 40) is placed inside a box, which is raised initially about 4 inches (10 cm) off the floor so the hatchlings can duck underneath the box to get warm and come out from under it to eat and drink. When the chicks appear to be more comfortable resting at the edge of the hover, or out from under it, raise the hover by 2 to 4 inches (5 to 10 cm) to improve air circulation and reduce the temperature at floor level. Typically, you will need to raise the hover about once a week.

A hanging hover is suspended by ropes or chains attached to the tops of the four corners. The four ropes or chains are brought together at the center and attached to a single rope going through a pulley in the ceiling, by means of which you can easily adjust the hover's height. The advantage of a hanging hover is that you can pull on the rope to raise the hover and look underneath for injured or ailing birds.

A standing hover has legs at the bottom four corners. The hover may be raised by setting the legs on bricks or blocks. A more secure alternative is to make the legs adjustable, as shown in the illustration opposite. A cardboard hover may be disposed of at the end of the brooding season and the adjustable legs saved for reuse.

Brooder Guard

The circular **brooder guard**, also known as a **draft shield** or **chick corral**, keeps baby birds from wandering far from the hover, reduces floor drafts, and eliminates corners where chicks tend to pile

up and can be smothered. The chick corral concept may also be used to eliminate corners in any square or rectangular brooder.

Brooder guards may be purchased by the roll, or you can easily make your own. The guard is basically a 12- to 18-inch (30–45 cm)-high fence made of corrugated cardboard, arranged in a circle around the hover, with the ends taped or stapled together. An alternative to a circular corral that still avoids corners is using six panels cut from sturdy cardboard and taped into a hexagon. Panels are added as needed for expansion.

Size and space. The chick corral must be big enough to accommodate the heat source, feeders and drinkers with surrounding space for access, plus room for all the chicks to get out from under the heater if they feel too warm. The diameter of the corral thus must be adjusted to the size of the hover. A 12-foot (3.75 m)-long guard will make a corral of about 4 feet (1.25 m) in diameter, the minimum size to accommodate a 2-foot (0.2 m)-square box hover. Six 2-foot (0.6 m)-long panels will give you a hexagonal corral of approximately the same size.

Initially, the corral should be no farther than about 2 feet (0.6 m) from the hover, in all directions, with feeders and drinkers placed at the hover's edge so no chicks can wander far and get chilled. After the first 2 or 3 days, when the chicks are eating and drinking well and have become more active, the corral may be expanded and the feeders and drinkers moved a little farther from the hover.

Adapting as chicks grow. In a week to 10 days, after the chicks have become familiar with the locations of heat, feed, and water, the chick corral may be removed. If it is still needed to reduce floor drafts, it must be enlarged to encompass a greater area to give the birds room to grow and must also be made taller to keep them from perching on top and coming down on the wrong side. In hot weather the cardboard chick corral might appropriately be replaced by a corral made from ¼-inch (6 mm) or ½-inch (13 mm) hardware cloth to improve ventilation while continuing to keep chicks from piling in corners or wandering away.

Housing for Tween-Age Birds

Depending on the size of your brooder and the number of chicks you brood, typically, they will rapidly outgrow their first brooder and must be moved to roomier quarters. If the weather is warm by then, and the birds are nearly feathered, they may no longer need additional heat, but they still require confinement for their protection and to keep each other warm.

One possibility is to move them to the coop where they will live but keep them confined indoors, perhaps opening the chicken-sized "pop hole" door to let them wander in an enclosed outdoor pen during nice weather. Screened windows or a screened door left open during the day will provide sufficient ventilation but should be closed at night to reduce drafts. Even if the coop is larger than they need at the time, they will huddle together to keep warm on cooler nights.

GROW PENS

An alternative is to move the chicks from the brooder to a grow pen — also called a grow-off unit or halfway house. It is a scaled-down chicken coop with a scaled-down chicken run or sunporch, two outdoor exercise area options. Broilers, in particular, get overheated easily as they gain bulk and therefore need to be moved out of a heated brooder into a grow pen by about four weeks of age, well before they are ready for harvest.

Likewise, cockerels separated from the pullets in a batch of straight-run chicks may be kept in a grow pen while you decide what to do with them. Replacement layers may also be raised in a grow pen, which need not have nest boxes because the hens should be moved to their final quarters by the time they start to lay.

This unit has an enclosed brooder at the right and a grow pen to the left, with a partition between the two right-hand doors to separate the two areas. The brooder section has a removable plywood floor and is wired for light and heat. A pop hole door in the partition provides access to the sun porch, which has hardware cloth on the floor and on the front and back walls.

In cold weather the front and/or back may be covered with clear plastic to retain warmth. In warm weather, the plastic is removed and one or more of the doors may be opened, and a ladder added, to let the birds out to enjoy sunshine and green grass.

Pasture Shelter

A portable shelter is a popular grow pen for broilers because they don't require permanent year-round housing. If the weather is warm enough — and presumably weather that's warm enough for grass to grow is warm enough for chicks — broilers may be put into a grazing area when they are as young as two weeks of age. If the weather is still cool, wait until the chicks are fully feathered before putting them out to forage.

Pullets, too, may be raised on forage. Letting them graze away from older birds gives them time to develop immunities through gradual exposure to the diseases in their environment. Even in a colder climate where the pullets must later be moved to winter housing, rearing them in a grassy area during spring and summer offers sunshine, fresh air, and green feed that combine to keep the birds healthy.

When moving chicks to a portable shelter, confine them for the first few days to the shelter or to a small, enclosed yard where they can become oriented before you give them freedom to roam. If it rains, especially a long, hard rain at night, go out and check your birds to make sure they aren't huddled outside or squatting in a puddle inside the shelter.

Even if you don't have grazing land, when chicks are at least two weeks old and the weather is warm and sunny, you can put them in a pen or wire-bottom cage on the lawn for a few hours each day, provided the lawn has not been sprayed with toxins. They'll need shade and water, a ½-inch (13 mm) wire-mesh guard around the perimeter so they can't wander away, and wire mesh over the top to keep out hawks and cats. For good sanitation and to provide fresh forage, put the pen in a new spot each day. If you put a grow pen in your garden, spade the soil in the previous location of the pen to combine the droppings with the soil.

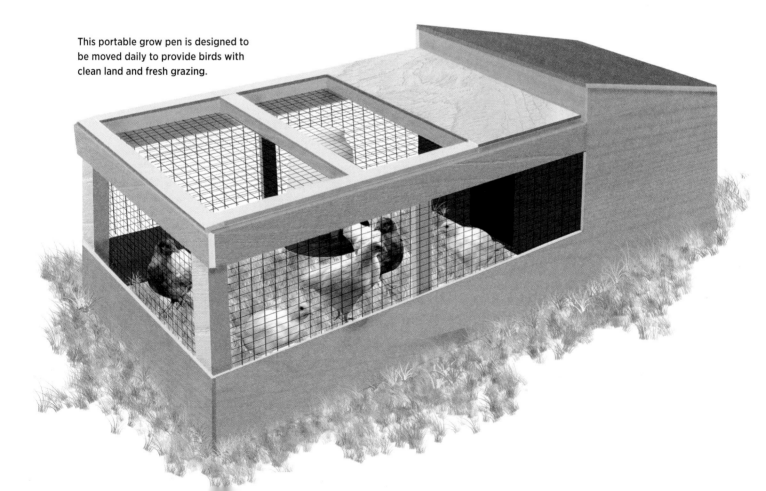

This portable grow pen is designed to be moved daily to provide birds with clean land and fresh grazing.

Providing Heat for the Little Ones

Homemade brooders are heated by one of two kinds of heat sources: incandescent and infrared. Incandescent heat is created by a source that produces light by being heated; in other words, a lightbulb of one sort or another. Infrared heat is generated by electromagnetic energy and does not involve light. Confusingly, the most common brooder heater is an infrared heat lamp, which falls in the middle by being primarily heat producing while also emitting light. Each option has advantages and disadvantages.

LIGHTBULBS

A lightbulb is a commonly used heat source for small batches of chicks. The types of lightbulbs suitable for brooding are incandescent and halogen. Compact fluorescents are not suitable, because they are more energy efficient and therefore produce less heat, although they may be used as auxiliary light in conjunction with an infrared heater (as discussed below).

Incandescent bulbs emit some 90 percent of their energy in the form of infrared radiation, or heat. Although standard incandescent bulbs are inexpensive to purchase, they don't last as long

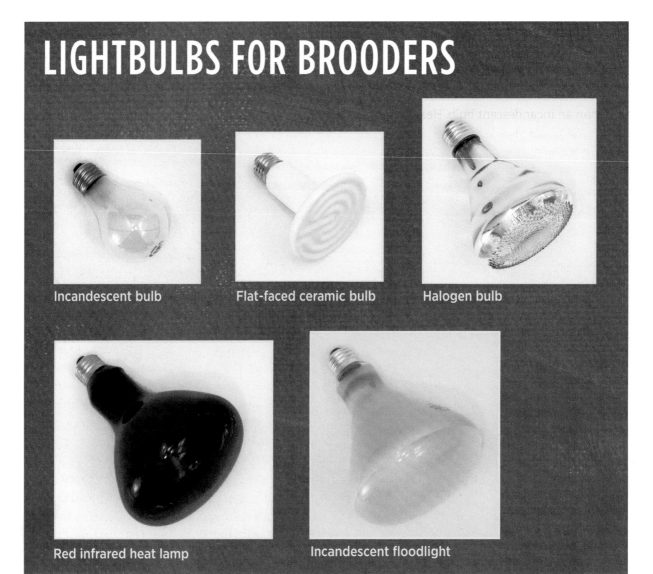

LIGHTBULBS FOR BROODERS

Incandescent bulb   Flat-faced ceramic bulb   Halogen bulb

Red infrared heat lamp   Incandescent floodlight

as other heat sources; the average life expectancy for continuous use is about 6 weeks. Incandescent floodlights are more expensive than standard bulbs but can last for nearly 4 months of continuous use. Floodlights designed for outdoor use are best for brooding purposes.

Halogen bulbs are incandescent bulbs that produce more light per unit of energy than normal incandescents but also produce more infrared radiation and therefore get a lot hotter. Yet a halogen bulb is at least 10 percent more energy efficient than a regular incandescent bulb and should last approximately as long as an incandescent floodlight. To get the maximum life out of a halogen bulb, do not touch the glass with bare hands (which leaves bulb-life-shortening fingerprints), avoid jolting the bulb while you're screwing it in or adjusting the fixture, and do not adjust a fixture while the bulb is hot.

An **infrared heat lamp**, sometimes called a brooding lamp, is basically an incandescent lightbulb that emits less light and more infrared radiation than an incandescent bulb. Heat lamps come with either red or clear bulbs. A white lamp may be expected to last for up to 30 weeks of continuous use. A red lamp is more expensive but should last about 6 weeks longer, and the red glow discourages chicks from picking at each other — as long as everything looks red, truly red things, such as bloodied areas left by feather picking, won't attract attention.

When purchasing an infrared heat lamp, opt for one rated for outdoor use or sold specifically for brooding chicks and other livestock. It will be slightly more expensive than a lamp designed for indoor use but is heavy duty and less susceptible to deterioration from the moisture in a brooding environment. To preserve long life, observe the same handling precautions as for a halogen bulb.

Watch More Than the Wattage

A bulb's wattage gives you only a general idea of its heat output. Theoretically, two bulbs with the same wattage will produce approximately the same amount of heat. In reality, two bulbs of identical wattage can be as much as 20°F (11°C) apart in heat output. Furthermore, the stated wattage — especially of imported bulbs — may be misrepresented. Whenever you put new birds into a brooder or change the bulb in the brooder's fixture, check the brooder often for the first hour or so to make sure your birds don't get chilled or overheated.

INFRARED HEATERS

An infrared heater is considerably more expensive than a light-emitting bulb but has the following advantages:

- It's longer lasting
- It's more energy efficient, therefore cheaper to operate
- It does not easily break or shatter
- It produces uniform, sunlike warmth without creating hot spots
- It emits no light, letting chicks rest at night

On the other hand, birds do need daytime light to find feed and water. Using an infrared heater therefore means you also need to provide light (as discussed below).

Ceramic bulbs and **panel heaters** are two types of infrared heaters typically used for brooding chicks on a small scale. As with lightbulbs, each of these options has advantages and disadvantages.

A ceramic bulb — also called a **ceramic infrared heat emitter** or an **Edison screw-base ceramic bulb** — looks something like a regular screw-in lightbulb and uses the same type of fixture. In many ways it is similar to an infrared heat lamp, except it is made of porcelain and of course does not emit light. A ceramic bulb can last for up to 5 years of continuous use, which means many years of seasonal brooding.

Flat-faced ceramic bulbs are more energy efficient and longer lasting than dome-shaped bulbs. Like light-emitting bulbs, ceramic bulbs come in different wattages. I use a 100-watt ceramic bulb in my big brooder.

FIND THE RIGHT BULB FIXTURE

Infrared lamps and bulbs get extremely hot and therefore require a fixture that can withstand the high heat and also deter people and birds from touching the hot bulb. Must-have features include:

- A porcelain socket that won't melt or catch fire (as a standard plastic socket might)
- A wire guard or safety cover to keep people and birds away from the hot bulb and to avoid starting a fire should the fixture fall against a wall or into bedding
- A secure method by which the fixture may be attached and adjusted

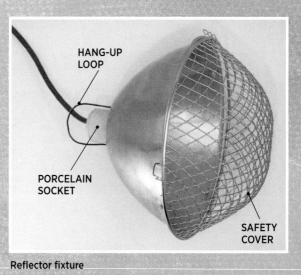

The fixture may have an open wire cage to prevent back heat from accumulating against the socket or (more commonly) a reflector that directs heat toward the brooding area. Some fixtures have a hang-up loop for easy suspension, and most have a clamp in addition to, or instead of, a hang-up loop. Securely hanging a fixture is safer than clamping it, as a clamp can be easily knocked loose by a careless person or a bird that's clumsily learning to fly. If the fixture has no hang-up loop, hang it from the clamp, or at least tie the clamp securely in place in case the clamp itself should slip loose. Never hang a lamp by its cord — that's unsafe.

Reflector fixture

Proper brooding fixtures are available from farm stores and pet supply outlets. They may hold one bulb or more than one, the latter having the advantage of not only accommodating more chicks but also continuing to provide heat should one of the bulbs burn out. An alternative to a multibulb fixture is two individual fixtures, to ensure that one bulb will remain functional should the other fail.

A brooding fixture may have a built-in **rheostat**, or dimmer switch, that allows you to adjust the heat output. Make sure the fixture you use has the same wattage rating as your bulb. A 250-watt infrared bulb, for instance, will put out more heat than can be handled safely by a fixture rated for up to 150 watts. The inverse — using a bulb with lower wattage than the fixture's rating — is not a problem.

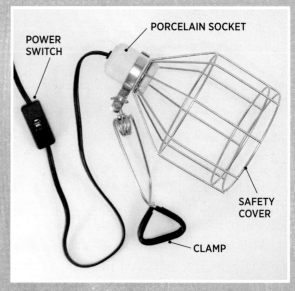

Wire cage fixture

PANEL HEATERS ARE SAFE AND EFFICIENT

An infrared radiant heat panel, designed for use with pets and livestock, is initially the most expensive option for brooding poultry. It is also, however, the most energy efficient and least expensive to operate and therefore over the long run is considerably cheaper than other options. And, when properly cared for, a panel heater may last indefinitely. It's the safest option as well because it does not get hot to the touch; it will not burn birds or their keepers and does not present a fire hazard. The sealed panels are easy to wipe free of chick dust and other brooding debris.

A radiant heat panel directs heat only beneath itself, giving chicks in a small brooding area the flexibility to move away to maintain their comfort level. And when the ambient temperature is warm enough that chicks barely need heat in the brooder, the panel "ceiling" may be turned off and lowered to trap and hold the body heat of chicks sleeping underneath, keeping them sufficiently warm.

Panels intended for brooding chicks come in various sizes to accommodate brooder size and number of chicks and are designed with built-in height adjustment. EcoGlow brand panels, for example, come with either a stand or legs for limited height adjustment, while Infratherm's Sweeter Heater panels hang from lightweight chains that offer unlimited height adjustment. The latter may therefore be lifted high enough to keep chicks from perching on top and are also usable off-season for such nonbrooding needs as preventing winter frostbite in grown chickens and warming an injured or ailing bird. EcoGlow panels are 12 volt and come with a transformer; Infratherm panels plug into a regular outlet and may be used with a rheostat to adjust heat output. I use Infratherm panels in six of my seven brooders.

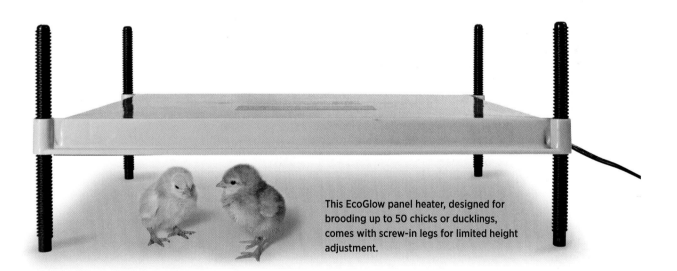

This EcoGlow panel heater, designed for brooding up to 50 chicks or ducklings, comes with screw-in legs for limited height adjustment.

Positioning the Heater

The best place to position the heat source in a roomy brooder, including an area brooder, is in the middle, with feeders and drinkers spaced around the outside edges. In a brooder with limited space, position the heater at one end to allow room for the feeder and drinker at the other end. Avoid positioning the heater directly above the feeder and drinker, which encourages chicks to bed down on or in the feed and water.

BROODER CHICK BODY LANGUAGE

You can easily tell how comfortable your chicks are by the way they arrange themselves in the brooder. Look for the following clues:

Chicks crowded close to the heater. The chicks are not warm enough. Typically they will peep shrilly. If they are chilled for long, they may develop sticky bottoms (see page 102). In an effort to get warm, they will pile on top of each other, smothering those unfortunate enough to be on the bottom. Even when chicks are cozy during the day, if the nights are cool, they may need extra heat overnight.

Chicks crowded away from the heater. The chicks are too hot. Typically, they will lift their little wings away from their bodies and pant. In an effort to get away from the heat, they will press into corners, piling on top of each other and smothering those on the bottom. If the brooder gets hot enough to raise their body temperature above 117°F (47°C), chicks will die.

Chicks crowded to one side of the brooder. These chicks may huddle as far away as they can from a cold draft, all facing the direction the draft is coming from. Typically, they will peep in distress, and some may die.

Chicks evenly distributed under heater. Chicks that are warm and cozy sleep spread out side by side, either directly under or at the edge of the heat source. Typically, these chicks make very little sound and barely move, except for the occasional chick adjusting its sleeping position.

Chicks active and evenly distributed throughout brooder. These chicks are perfectly comfortable. Typically, they emit musical sounds of contentment and keep busy eating, drinking, and wandering freely throughout the brooding area while they explore their little world.

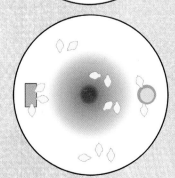

CONTROLLING HEAT

How much heat chicks need, and for how long, are not questions with easy answers. So many variables are involved: the species raised, the breed, their numbers, their rate of growth, the size of the brooder, the room temperature in which the brooder is located, and the method of providing heat. Compared to chicks, poults are quicker to chill, while keets, ducklings, and goslings do not chill as readily.

People, including sometimes me, tend to err on the side of providing more heat than their chicks need for longer than they need it. That's not doing the birds any favors (it's like heating a chicken coop, which may be more comfortable for the chicken keeper but in most cases is detrimental to the chickens). Regularly throttling back on the brooder heat is the better option.

Theoretically, brooder temperature is measured with a thermometer placed 2 inches (5 cm) above the brooder floor (and at the outer edge of a hover), but using a thermometer is impractical, and anyway, you shouldn't need one. Just watch the chicks, and adjust the temperature according to their body language.

Tweaking Temperature

You can adjust brooder heat to maintain chick comfort in several different ways. As chicks grow — producing more body heat as they gain weight, while requiring less heat input as they feather out — brooder heat may be adjusted through one or more of the following methods.

Raise the heater to increase the distance between the heat source and the chicks. The best way to do this is to hang the heater by an adjustable chain. Start a bulb heater about 18 inches (45 cm) above the chicks or a hover or panel heater about 4 to 6 inches (10–15 cm) above the brooder floor or just high enough for the chicks to walk under without bumping their heads. As the chicks grow raise the heater by about 3 inches (7.5 cm) each week, or as much as necessary for it to remain well above their heads and still maintain their comfort level. Do not remove the protective safety cage around a bulb heater. It prevents burns to birds that clumsily try to fly about. And a panel heater isn't that hot; my chicks, keets, and poults frequently sleep sprawled out on top of the panel.

Lightweight chain is ideal for raising and lowering heaters. It is commonly attached by means of S hooks, but after spending way too much time looking for lost S hooks in chick bedding, I switched all my S hooks to key rings.

Reduce heat output by switching to lower-wattage bulbs or smaller heater panels. Bulb heaters come in a variety of wattages. For halogen and incandescent bulbs, 100 is the wattage most commonly used for brooding up to two dozen birds, although a lower wattage may be needed in a tightly confining brooder, such as one made from a plastic storage tote or for a brooder located in a warm room.

Infrared lamps are available in wattages up to 250, the most commonly used size, although in a confined situation 250 watts may provide too much heat. At an average room temperature of 50°F (10°C), a 125-watt lamp should be sufficient for about three dozen chicks. If the room temperature is warmer, the same lamp can accommodate more chicks (five more for every 10°F [5.5°C] increase); if cooler, fewer chicks (five fewer for every 10°F decrease). By expanding the brooding area and using a 250-watt lamp, you can double the number of chicks brooded.

Rule of Thumb

Start the brooding temperature at 90 to 95°F (32–35°C) and reduce it approximately 5°F (3°C) each week until the brooder temperature is the same as the ambient temperature, or about 70°F (21°C), whichever comes first. This means that in hot weather, or in a well-heated room, the brooder may require little or no additional heat. Within the chicks' comfort zone, the more quickly you reduce the heat level, the more quickly the chicks will feather out and the sooner they can do without auxiliary heat.

Ceramic bulb heaters also range up to 250 watts and, watt for watt, put out a similar amount of heat as a heat lamp. If an infrared lamp or ceramic bulb makes your chicks too hot, switch to a lower wattage or to an incandescent bulb.

A panel heater functions much like a hover and therefore should be large enough for all the chicks to sleep underneath at the same time. As a general rule, allow about 4 square inches (25 sq cm) of panel per chick. In my tote brooders I hang either an 11-inch-by-11-inch (28 × 28 cm) Infratherm panel or an 11-inch-by-16-inch (28 × 40 cm) panel. The smaller panel works best when the room temperature is warm and the tote is at maximum capacity (about 30 chicks) because that number of chicks in the confinement of a tote generates a lot of body heat. The larger panel works best when the room is cool or I don't have many chicks to brood.

Reduce the heat output by using a rheostat. When the heater panel in one of my tote brooders is raised as high as it will go under the plastic and hardware cloth tote cover, I adjust the heat output by plugging it into a rheostat. Most ready-made brooders have a built-in thermostat for heat adjustment.

Improve ventilation. As the weather warms and chicks feather out, seemingly overnight they can go from needing heat to needing to be cooled off through better air circulation. Chicks in a hot, poorly ventilated brooder may die after just 3 hours at 106°F (41°C). Some may die even after their comfort level has improved, and survivors may not grow well. As the temperature rises and chicks start acting uncomfortable, open windows to improve ventilation, while taking care to avoid letting a draft blow across the brooder floor.

Increase the amount of available living space so the chicks can adjust their own comfort level by moving away from the heat and by spreading apart from each other. I have a series of 3-foot-by-4-foot (0.9 × 1.25 m) brooders built along one side of a stall in my barn, each heated with an 11-inch-by-30-inch (28 × 76 cm) Infratherm panel. When chicks get crowded in one of the brooders, I open the front of the brooder and move their feeder and drinker into the stall. The chicks are then free to run around in the stall or to go back inside the brooder when they want to warm up or are ready to bed down for the night.

Two ways to control heat are by increasing the amount of living space (get a bigger box) and by raising the heat source (add a second, smaller box on top to create a heater tower).

How Long Should I Provide Heat?

Exactly how long growing birds need to be kept warm depends on the ambient temperature, the number of birds for the size of the brooder, and the breed. In warm weather they may need heat for 3 weeks or less. In cold weather they may need to be heated for as long as 6 weeks, until they grow enough feathers to keep themselves fully warm.

The more chicks you brood together, the more body heat they will generate to keep each other warm and the sooner you can reduce the auxiliary heat. Fast-growing broilers also need a shorter brooding period than other chicks, since they overheat more readily than do most other breeds. If they are overheated for long, they will eat less and grow more slowly. In warm weather broilers may need heat for no more than 2 weeks; in cooler weather they should not need auxiliary heat for more than 4 weeks.

Chicks of all breeds suffer more when they're too hot than when they're a little on the cool side. Chicks raised through the brooding period, with the heat reduced gradually but a little more rapidly than usual, will feather out more quickly, reducing the amount of time they need external heat.

The best way to tell when growing birds no longer need heat is to watch their body language. When they spend most of their time as far from the heater as they can get, and especially if they pant or lift their wings, it's time to turn off the heat.

Controlling Light

Whether or not the heat source also produces light, chicks do need light to help them find feed and water. Furnish continuous light for the first 48 hours while they are getting oriented to the brooder and to the feeders and drinkers. Thereafter, the light may be adjusted or controlled to provide some period of darkness daily.

Continuous light is not natural for birds. A natural period of darkness every 24 hours gives them a chance to rest. Heaters that do not emit light are therefore better for brooding purposes than heat bulbs that light the brooder 24/7. Compact fluorescent lights work well in combination with non-light-emitting heaters, because they do not produce additional heat that can make the brooder too hot. Use a 40- or 60-watt bulb for the first few days, then change to 15 or 25 watts as the chicks grow. Reducing the wattage not only reduces the heat output but also reduces the light level to keep chicks calmer and less likely to start picking as they grow.

If the brooding area gets plenty of natural light through windows or is situated in a well-lighted room, it may get sufficient light without an additional brooder light. Watch to see that the chicks continue to eat and drink at the normal rate.

If the brooder is lit by an area or brooder light, turn the light off when you go to bed at night and turn it back on in the morning. Or put it on a timer that allows at least 8 hours of darkness.

Of course, if the source of light is also the source of heat, you can't turn it off overnight. But at least turn it off for half an hour during each 24-hour period — obviously, not during the coolest hours — so the little birds will learn not to panic and pile up later when lights are turned off at night or in the event of a power failure.

Light affects the growth rate of chicks, so never keep them entirely in the dark. Even if you dim the light to prevent picking, leave enough light for you to see what's going on in the brooder, and it'll be sufficient for the chicks. A rule of thumb is to provide at least enough light to allow you to read a newspaper.

SPACE NEEDS QUICKLY CHANGE

Hatchlings remind me of popcorn — you put a few small kernels into a popper and before you know it they expand to fill the space. And so it is with chicks in a brooder. One day they have plenty of room in which to roam, and seemingly the next day the brooder is wall-to-wall chicks.

The rate at which chicks grow varies from breed to breed. Hybrid broilers are bred for excessively rapid growth, and commercial layers are bred for early maturity. Bantams and most of the old-time breeds grow more slowly. Separating the pullets from the cockerels as soon as you can identify them helps both groups grow more steadily.

For their first few days of life, hatchlings spend a lot of time sleeping and therefore don't need much room to move around in. But as they get bigger and more active, they increasingly need more space, both for sanitary reasons and to prevent boredom that leads to picking at each other. They also need enough space to be able to get out from under the heater if they feel too warm. On the other hand, more isn't necessarily better — given too much space during cold weather, young birds have trouble staying warm.

If you start chicks in a cardboard box, plastic tote, or other closely confined brooder, giving them more space as they grow means either dividing them up into two or more boxes or periodically moving the entire batch to larger quarters. If you start chicks in an area brooder, giving them more room is simply a matter of expanding the chick corral until it is no longer needed.

The minimum space to begin with is about 6 square inches (40 sq cm) of floor space per chick. Bantams and light breeds can get by with as little as 4 square inches (25 sq cm), while broilers and the really big breeds need more like 8 (50 sq cm).

Naturally, if you start with the minimum brooder size, you'll have to increase the brooding area sooner than if you provide a larger brooder to begin with.

Base the size of your brooding and growing space on common sense and observation rather than on a meticulous measuring of the floor space. Observation tells you chicks are overdue for expanded living quarters if:

- They have little room to move and exercise or to spread out for sleep
- They dirty the brooder floor faster than you can keep it reasonably clean — droppings pack on the floor, manure balls stick to feet, or you can smell ammonia
- They run out of feed or water between feedings, indicating the need for a larger area to accommodate more or larger feeders and drinkers

A move to unfamiliar housing is a frightening experience for chicks, and chicks that are frightened tend to pile together and smother one another. For the first few nights after moving them, provide dim lighting and check often to make sure they are okay. Moving their old feeders and drinkers to the new location also helps by bringing along things that are familiar to them.

Although you needn't initially acquire all the necessary equipment for your growing birds, they will rapidly outgrow their first feeders and drinkers. You must soon replace these with larger units that provide better access and hold more feed and water for maturing birds. Similarly, the most suitable bedding options change as the birds become strong and active and as they generate greater quantities of moisture and manure.

Infratherm Sweeter Heater panels come in several sizes and are hung from lightweight chains to offer unlimited height adjustment as hatchlings grow.

3

MANAGING WATER, FEED, AND BEDDING

Your brooder is selected and your hatchlings are scheduled to arrive on your doorstep soon. Now's the time to get everything set up and ready for them. Preparing the brooder in advance gives you time to round up anything you might have forgotten, including drinkers, feeders, starter ration, or bedding. At least a day before their arrival, warm up the brooder so it'll be nice and cozy for your baby birds.

Providing Water

Hatchlings can go without water for their first 48 hours of life, but the sooner they drink, the less stressed they will be and the better they will grow. A chick's body needs water for all life processes, including digestion, metabolism, and respiration. Water helps regulate body temperature by taking up and giving off heat and also carries away body wastes.

THAT FIRST IMPORTANT SIP

A hatchling's first drink should be at brooder temperature. Fill and place the drinker at the same time you turn on the heat, and the water will warm up to the right temperature. If you forget to fill the drinker before the chicks arrive, use warm (not hot) water from the tap. Drip some of the water onto the inside of your wrist; if it feels neither warm nor cold, it's the right temperature. Hatchlings that drink cold water can get chilled, especially if they arrive thirsty.

To make sure each bird finds the water and starts drinking right away, a common practice is to gently push its head toward the water, thereby dipping its beak into the drinker and watch that it swallows before releasing the bird into the brooder. When you do your own hatching, it's probably not necessary. Hatchlings are naturally attracted to shiny things, including the surface of water, and will automatically peck at it. Once they get their beaks wet, they will start drinking.

Hatchlings that have been in transit a day or two may arrive dehydrated and disoriented. Birds

A 1-quart (0.9 L) drinker holds enough water for up to 25 hatchlings.

that have been left in the incubator too long after they hatch also get dehydrated. Dipping their beaks into warm water helps them quickly learn where the drinker is and encourages them to drink more, ensuring timely rehydration. After their beaks have been dipped, some of the birds may start drinking right away, others may not. That's okay. As long as one chick drinks, the others soon follow the leader.

While land fowl may be slow to drink, waterfowl babies — particularly ducklings — can be a little too eager. When offering first water to ducklings, watch to make sure they don't overdo it. A dehydrated duckling that drinks too much at once can go into shock. If ducklings seem too eager to fill up on water, let them have access to the drinker for 10 to 15 minutes, then remove the water for 15 to 30 minutes. After they have had four of these sessions at the water fount, with time to rest after each session, they should slow down enough for you to permanently return the drinker to the brooder.

Talking Fowl

drinker. Also called waterer or fount; a container from which birds drink water

feeder. A container from which birds eat their daily rations

trough. A long, narrow drinker or feeder

tube feeder. A cylindrical feeder into which rations are poured at the top and birds eat from a circular pan at the bottom

WATER QUALITY

Give your hatchlings only clean, clear water you would drink yourself. Chicks that are fresh out of the incubator should need no additives to remain healthy, although many chicken keepers feel that a health-booster solution helps hatchlings cope with the stress resulting from their transformation from embryo to chick.

Being shipped cross-country increases the stress level — the more so the longer the birds have been in transit. A booster solution of sugar furnishes extra calories that give chicks a spurt of energy and, because they like the sweetness, encourages them to drink more to speed up rehydration. Stir up to ½ cup (118 mL) of table sugar into each quart of water (120 mL sugar per 1 L water) for the first day or two, after which revert to plain water. Too much sugar can cause loose droppings.

Additives

All sorts of health-booster solutions, consisting primarily of vitamins and electrolytes, are promoted as necessary to help chicks overcome shipping stress and give them the best start in life. Throughout the years I have brooded shipped chicks with and without such solutions and can't see any difference. The chicks I hatch myself thrive on nothing more in their drinker than clean fresh tap water, changed as often as necessary to keep it clean and fresh.

How often the water must be changed depends on how many birds you are brooding, the temperature, their drinking habits, and how much your drinkers hold in relation to the number of birds. We rinse clean and refill drinkers twice a day, morning and evening. Doing so ensures the babies have plenty of water and also removes any accumulated sludge caused by feed or bedding falling into the drinker. As the baby birds grow, and especially when the weather becomes warmer than the brooding temperature, we also check drinkers in the middle of the day in case the chicks have drunk all the water or knocked one over.

Marbles or clean gravel placed in the drinker's rim for the first few days will keep keets and tiny bantams from drowning in the water.

SUITABLE DRINKERS

Hatchlings must have access to fresh, clean water at all times. Initially, a 1-quart (0.9 L) drinker furnishes sufficient water for up to 25 chicks. A standard chick drinker consists of a 1-quart container, such as a canning jar, screwed into a round plastic or galvanized steel base. A plastic base cracks over time, and a metal one breaks away from the portion that screws onto the jar, so it pays to keep a few extras on hand. If you intend to put any type of additive into the drinking water, don't take a chance that it may interact with a metal base — opt for plastic. Drinker bases, as well as entire drinkers consisting of a plastic base and a plastic jar, are available from most feed stores and poultry-supply catalogs.

Don't be tempted to cut corners by furnishing water in an open dish or saucer. Chicks will walk in it, filling the dish with bedding and droppings that spread disease. They'll be more likely to get wet and chilled. And some chicks may drown.

Drowning is generally not an issue when chicks have a proper drinker unless they are so crowded they can't spread out to sleep without some chicks getting their heads in water. Keets and tiny bantams may have trouble with drinkers designed for larger chicks, in which case you can put marbles or clean gravel in the water for the first few days until the birds get big enough not to fall into the water. Some suppliers offer a drown-proof small bird drinker with a narrower

Trouble-Free Water Nipples

Inexpensive water nipples may be used to make drinkers that are suitable for all ages and all species of poultry. If they are properly installed so they don't leak, they are virtually trouble free. Nipples don't spill over and dampen the bedding, they keep bedding and poop out of the drinking water, and they don't need to be scrubbed out every day. Just check each nipple daily to make sure it isn't leaking or clogged (tap it with a finger to make sure water comes out).

Nipples are easy to use. They may be installed along a PVC water line that is gravity fed from a manually (or rain-) filled 5-gallon (19 L) bucket or 55-gallon drum. Or the nipples may be attached to the tops or bottoms of water or soda bottles, the bottom of a milk jug, or the bottom of a 2-, 3-, or 5-gallon (7.5, 11.5, or 19 L) bucket. Just drill a hole of suitable size and insert the nipple so it hangs down vertically.

Nipples come in two styles: push in and screw in.

For the push-in style, drill a 27/64-inch (1 cm) hole, push in a rubber grommet, then push in the nipple.

For the screw-in style, drill an 11/32-inch (0.9 cm) hole into thick plastic, or a 5/16-inch (0.75 cm) hole into thin plastic, such as a bottle cap; wrap the nipple's screw thread with silicon pipe-thread tape; and tightly screw the nipple into the hole.

Nipple manufacturers recommend one nipple per three birds. In practice you can get by with one nipple for up to six birds. As with any drinker the height must be adjusted as chicks grow; keep the nipple at, or just above, eye level. An Internet key word search for "chicken water nipples" will net you lots of drinker design ideas, as well as important information on how to establish appropriate water pressure for in-line nipples.

Water nipples are inexpensive, sanitary, and virtually trouble free.

drinking area to keep the smallest hatchlings from drowning.

Within a week or so, the chicks outgrow a 1-quart watering jar. You'll know the time has come when you find yourself filling drinkers increasingly more often to keep your chicks in water at all times. Another sign that a waterer is too small is finding chicks perched on top, a habit that lets droppings get into the water. The time for changing drinkers has long since passed when you find the quart jar overturned and spilled due to boisterous play in the brooder.

Bigger Drinkers

When it's time to change to larger drinkers, you have several choices. Regardless of the design, a good waterer:

- Is the correct size for the flock's size and age — chicks should neither use up the available water quickly nor be able to tip the fount over
- Has a basin of the correct height — a chick drinks more and spills less when the water level is between its eye and the height of its back
- Isn't easy to roost over or step into — droppings plus drinking water equal a sure formula for disease
- Is easy to clean — a fount that's hard to clean won't be sanitized as often as it should be
- Does not leak — leaky drinkers not only run out fast but create damp conditions that promote disease

Bell-style drinkers are similar to the quart-jar setup, only in larger versions. Depending on the number of birds you're brooding and the amount of space they have, you might first switch from quart jars to 1-gallon (3.75 L) plastic drinkers, then move up to 3-gallon (11.5 L) or 5-gallon (19 L) metal founts if they empty the smaller fount too fast, such as when the weather turns warm.

Automatic drinkers ensure your chicks won't run out of water when no one is around to check and refill founts, and they are a great time saver for brooding large numbers of chicks. Automatic founts come in several different styles, including trough, cup, and nipple. A trough may be hooked up to a faucet with a garden hose, and the water level is controlled by means of a float valve. The trough should have a wire guard that lets birds get only their heads through, and the float valve should have a protective shield to keep birds (especially goslings) from fiddling with it until it falls apart. With this style of drinker it's a good idea to use more than one — in case one fails to work properly.

Cups and nipples are designed to work under low pressure, no more than 5 pounds per square inch (psi). They therefore require a gravity flow system, or if you want to hook them up to your main water line, you'll need a pressure regulator. Cups tend to accumulate feed and therefore must be cleaned frequently. Nipples are the most sanitary and trouble-free option. Automatic drinkers must be checked daily to make sure they aren't clogged or leaking.

An automatic trough fount connected by garden hose to a faucet is ideal for brooding waterfowl.

EASY-TO-FIND DRINKING STATIONS

A hatchling that loses 10 percent of its body water through dehydration and excretion will experience serious physical disorders. A chick that can't find water won't grow at the normal rate, develops

A drinker hung on a chain may be easily adjusted as chicks grow. A hardware cloth platform over a catch pan, placed beneath the drinker, helps confine drips.

a bluish beak, and stops peeping. If the loss of body water reaches 20 percent, the chick will die. The most common brooding situation in which chicks cannot find water is in an area brooder in which the drinkers are initially placed too far from the heat source. After 4 days without finding water, chicks start dying, and deaths continue until the sixth day. Any chicks remaining after that were the lucky ones — they managed to find a drinker.

To avoid such a catastrophe, initially place drinkers no more than 24 inches (60 cm) from the heat source. Later, as your chicks grow into expanded housing, make sure they never have to travel more than 10 feet (3 m) to get a drink. Whenever you change to a different style of drinker, leave the old ones in place for a few days until the birds get used to the new ones. Observe this stress-reducing precaution whether or not they are moved at the same time their drinkers are upgraded.

During the first days of brooding, place each drinker on a flat surface, such as a small square of plywood or a large ceramic tile. The purpose of this platform is to steady the drinker to reduce the chance it will tip and leak. After the first week, place each fount over a platform created from a 1×2 wooden frame covered with ½-inch (13 mm)-mesh hardware cloth. Drips will fall below the mesh, where chicks can't walk or peck in damp

Spacing Feeders and Drinkers

Some chicks start life quite active and curious; others aren't adventuresome at all. For the first day or two, place feeders and drinkers close to the heater, where all chicks can easily find them — but not directly under the heat, where the chicks may crowd against the feeders and drinkers while they sleep. As soon as the birds are eating and drinking well, move feeders and drinkers farther out to give the birds more space to rest near the edge of the heater.

Put as much distance between the feeder and drinker as the size of your brooder will allow. Chicks will drop feed from their beaks, as well as toss or scratch feed from the feeder, into the drinking water, creating an unhealthful sludge that can ferment if left too long in the warm brooder environment. Spacing the feeders and drinkers as far apart as possible reduces the amount of sludge that accumulates in the drinkers, and traveling between the feeder and drinker is good exercise for strengthening little legs.

bedding, and raising the drinker onto a platform helps keep bedding out of the water.

As your chicks grow, adjust the height of the drinker to keep the water level between their eyes and their backs. The easiest way to adjust a drinker's height is to hang it from a rope or chain, which might be attached to the brooder ceiling or to a large shelf bracket that's attached to the brooder wall. Lamp chain with 1-inch (2.5 cm) links is ideal for this purpose, because it lets you easily raise (or lower) the drinker in 1-inch increments. Many people use an S hook to attach the drinker to the chain, but I much prefer a snap hook because it won't fall off and get lost in the bedding. Rowdy chicks in the brooder can cause the drinker to swing and spill, so leave the hardware cloth platform below the drinker even if the drinker does not rest directly on the platform.

DRINKERS FOR DUCKLINGS AND GOSLINGS

Waterfowl use a lot more water than land fowl, because they must drink more to remain healthy and because they splash a lot while playing in the drinker. They also drop feed from their bills into the drinker, sludging up the water faster than land fowl do.

Ducklings and goslings are notorious for quickly turning their brooder and themselves into a soggy mess. Placing the drinker in a pie tin or similarly shallow container helps keep things under control. Better still is to place the drinker on a hardware cloth platform with a shallow catch basin underneath. After only a few days, waterfowl outgrow a regular chick drinker. A trough fount may then be used; it holds enough water for the birds to dip their whole heads into and has a wire guard to prevent them from swimming in the trough.

Homemade drinkers suitable for ducklings and goslings may be fashioned from plastic food containers of appropriate size. Margarine or cottage cheese tubs, or other similarly shallow containers, work well for babies. Cut a hole into the lid along one edge, just a little larger than the size of a bird's head, through which they can drink and wash their faces but cannot climb in for a swim. A milk jug, an ice cream tub, or a plastic coffee canister may be used for growing birds. As they grow they tend to push on or try to step on the drinker, which may tip over and spill unless it is anchored down. Put clean rocks into the drinker to make it bottom heavy and therefore more stable, or tie the drinker to the side of the brooder. To initially encourage your baby waterfowl to use your homemade drinkers, sprinkle a little finely chopped lettuce on the water to attract their attention.

Brooding Challenges, by Species

Keets	Easiest of all; few hardiness issues; they fly early and therefore need a secure brooder to keep them safely confined
Chicks (other than bantams and broilers)	Easy to brood; hardiness issues vary with breed; faster growing breeds require frequent increases in available living space
Bantam chicks	Easy to brood; hardiness issues vary with breed; the smallest breeds require drinker management to prevent drowning
Broilers	Rapid growth may need to be controlled to avoid lameness issues; must be carefully monitored to avoid overheating
Poults	May need help initially finding feed and water; early deaths are not uncommon; they perch early and will roost on anything they can reach
Ducklings	Few hardiness issues; controlling moisture is the chief concern
Goslings	Few hardiness issues; controlling moisture is the chief concern; slow rate of growth means a longer brooding period than for other species

HOMEMADE WATERFOWL DRINKERS

A milk jug with a hole cut into the side works well for baby waterfowl; as they grow taller, cut a new jug so the hole is at the top.

A shallow food tub makes an inexpensive drinker for ducklings and goslings.

A gallon-size plastic storage container with a slot cut along one side makes a handy drinker.

Algae Busters

Green scum appearing in drinkers, particularly those made of glass or plastic, is algae. Three elements are needed for algae to develop: water, light, and nutrients. Some algae are toxic, and no easy method is known for determining whether or not specific algae will produce toxins. The following methods may be used to help control algae:

- Rinse out drinkers as often as necessary to remove nutrient-rich feed sludge
- Thoroughly scrub out drinkers at least once a week — twice weekly is better
- Keep drinkers filled with fresh, clean water
- If tap water isn't pure, either filter it or boil and cool it
- Place drinkers away from sunlight
- Add 1 to 2 tablespoons of apple cider vinegar per gallon of water
- Put a thin slice of lemon into each drinker
- Put a piece of copper (such as a water-pipe fitting — but not pennies, which are mostly zinc) into each drinker

Galvanized steel drinkers keep out light and therefore are less susceptible to algae growth than plastic or glass drinkers. Do not use vinegar, lemon, or copper with metal drinkers, which can result in a chemical reaction.

Providing Feed

Newly hatched poultry come equipped with yolk reserves that continue providing nutrients for many hours after they hatch. It's nature's way of allowing the early hatchers to remain in the nest where they are safe and warm until all the stragglers have hatched. Shipping hatchlings by mail takes advantage of these yolk reserves, which are pretty well depleted by the time the birds arrive. Feed them within 2 to 3 hours *after* they have had their first drink, to make sure they are well hydrated before they start eating.

Even though hatchlings can survive for up to 4 days without eating, they experience less stress if they eat on the first day after they hatch. Chicks you hatch yourself, and that eat soon after being placed in the brooder, therefore have a better start in life than shipped chicks. Still, as with shipped chicks, make sure they are drinking before they start eating. Chicks tend to digest better if they get a good dose of water before getting a belly full of dry feed.

String Alert

When opening a sack of starter ration, keep track of the strings pulled from the end of the sack and take care to dispose of them properly. Watch not only for the cotton string used to sew the end of the bag together, but also for any plastic strips that might come loose from the polywoven sack. I have seen a hatchling die after swallowing a plastic string that passed through its digestive system, and I have seen strings get wrapped around a hatchling's feet or leg, or even its tongue. Prevent injury or death to your baby birds by making sure feed sack strings and plastic strips don't get into their feeder or brooder.

FEED OPTIONS

The easiest way to ensure that hatchlings get all the nutrients they need is to feed them a commercial starter ration, which contains a mixture of grains, protein, vitamins, and minerals. Formulating your own starter ration is tricky business, requiring knowledge of poultry nutrition for your chosen species, as well as the nutritional values of feedstuffs available to you. Furthermore, for just a few backyard birds, obtaining all the necessary ingredients, and grinding and mixing them into an appropriate ration, is simply not practical. If you're tempted to learn about mixing your own rations, do so after you become experienced at raising poultry, have a good handle on their needs, and can readily recognize deficiencies so you can make rapid corrections.

Chicken starter ration ("chick starter") is available in most locations. Starter rations are higher in protein and lower in calories than rations designed for older poultry. Never feed layer ration to babies, even as an emergency measure if you run out of starter; the higher calcium content of the layer ration can seriously damage young kidneys. If you do run out of starter, or you forget to pick some up before your hatchlings arrive, you can make an emergency starter ration by cracking scratch grains in the blender or, if you have no scratch, by running a little uncooked oatmeal through the blender and mixing it 50/50 with cornmeal. Grains are high in calories and low in the protein, vitamins, and minerals needed for good growth and health, so don't use this mixture any longer than a day or two.

If you have extra eggs on hand, mashed hard-boiled or scrambled egg makes an excellent starter ration. In the old days before commercial rations were available, farmers typically started their chicks on mashed boiled eggs. After the first few days, chicks were fed oatmeal and cracked grains until they got big enough to forage for themselves. Baby birds that are fed cracked grains also need grit, provided in a separate feeder. If your local farm store doesn't carry chick grit, a suitable substitute is cage-bird grit from a pet store or clean sand from a builder's supply.

Feeding Keets, Poults and Waterfowl Correctly

Depending on where you live, your local farm store may carry (or be willing to order) rations specific to the needs of poults and waterfowl. Turkey starter contains more protein than chick starter and is suitable for keets and waterfowl as well. Where turkey starter is not available, game-bird starter is a good alternative. Higher protein promotes maximum growth, which is desirable if you're raising poultry for the freezer.

I have raised chicks, poults, keets, ducklings, and goslings, all on starter intended for chickens without ever having a problem, but I don't feed for maximum growth, and I don't use medicated feed for any of my poultry.

Medicated feed that is not formulated specifically for waterfowl should never be fed to ducklings and goslings, both because they do not need the same medications as land fowl and because they do not eat the same quantities of feed as land

fowl and therefore may overdose on a medication not intended for them.

The only other serious issue involved in feeding land-fowl starter to waterfowl is that ducklings and goslings need more niacin than other poultry and therefore can experience niacin deficiency if fed solely on chick starter. Niacin deficiency easily may be avoided by adding 8 ounces (0.25 L) of brewer's yeast to each 10 pounds (4.5 kg) of starter. Brewer's yeast is available at supermarkets and health-food stores, although livestock-grade brewer's yeast from a farm store is cheaper.

Feeding Broiler Chicks

In areas where poultry is big business, farm stores may offer different feeds for poultry destined for the freezer compared to those that will become layers or breeders. Broilers, especially Cornish-cross strains, grow faster than other chicks and therefore need more protein. Broiler ration is usually 20 percent protein or more, compared to regular starter, which usually contains 18 to 20 percent protein. You may also find available a separate grower or finisher ration for broilers and a grower or developer ration for layers and breeders. Switch

Medicated Starter for Some Flocks

Some brands of commercial chick, turkey, and game-bird starter are medicated with a coccidiostat to prevent coccidiosis. Whether or not you need to feed medicated starter depends primarily on your management style. Use medicated starter for land fowl if:

- You brood chicks or poults in warm, humid weather
- You brood large quantities of poultry at a time
- You keep chicks in the same brooder for more than 3 weeks
- You brood one batch of chicks after another in the same brooder

You shouldn't need medicated starter if: you brood in late winter or early spring (before warm weather allows coccidia and other pathogens to thrive); you brood on a noncommercial scale; you keep your chicks on clean, dry litter by carefully managing sanitation and the drinker, and by providing enough ventilation; and your chicks always have fresh, clean drinking water. On the other hand, if you're raising your first-ever chicks, using medicated starter gives you one less thing to worry about while you work through your learning curve.

When raising poultry for the freezer, discontinue the medicated ration prior to butchering, following the withdrawal time period recommended on the label (usually *one* week) so that the drug is out of the bird's system before being consumed.

Remember: *Never feed medicated land-fowl starter to waterfowl.*

from one ration to another as indicated on the labels, gradually making the switch by combining the old ration with greater and greater amounts of the new ration to avoid problems related to digestive upset.

In many parts of the country, farm stores carry only one all-purpose starter or starter-grower ration. If you can find only a single ration, continue using it until broilers reach butchering age and layers and breeders are ready for the switch to lay ration.

Treats

Until birds are big enough to forage outdoors, they will enjoy tiny bits of fresh fruits and veggies, such as dark green lettuce, pieces of grapes or apple, and bean or alfalfa sprouts, all chopped into baby-bird-size pieces. Another treat they'll relish is dry baby cereal — barley, oat, wheat, or rice — from the supermarket. Offer treats only in small amounts, no more than they will gobble down in 15 minutes. For sanitary reasons, feed treats in a trough, on a tray, or on a piece of cardboard, rather than tossing them into the bedding.

Probiotics and Prebiotics

The small intestine of a healthy bird (and also a healthy human) is populated with a number of beneficial bacteria and yeasts, called intestinal flora or microflora, which aid digestion and stimulate the immune system. If for some reason these good guys get overrun by disease-causing microbes, the bad guys take over and cause an intestinal disease.

A hatchling acquires some microflora through the egg and gains more from the environment but may not develop a population of microflora fast enough to ward off disease-causing microbes. Chicks raised in a brooder, for instance, acquire beneficial gut flora more slowly than chicks raised under a hen, and those raised entirely on wire have even less opportunity to acquire microflora. Furthermore, extreme stress can affect the microflora population, and the use of antibiotics and other antimicrobials kills both disease-causing microbes and beneficial microflora alike.

Probiotics are live microflora introduced as a supplement that is either dissolved in water or added to feed. The idea is to boost immunity by giving chicks an early dose of the same gut flora that will eventually colonize their intestines. Probiotics improve intestinal balance through the process of competitive exclusion — meaning whoever gets there first wins.

The probiotic microorganisms work in three ways: by attaching to the intestinal wall to form a barrier that blocks out bad guys; by producing antibacterial compounds and enzymes; and by stimulating the bird's immune system. Packaged probiotic supplements especially formulated for poultry are available from many poultry suppliers. Natural sources include grains, meats, and fermented milk products, such as yogurt and kefir.

Prebiotics are nondigestible carbohydrate fibers that stimulate the growth and activity of gut flora and are commonly used in conjunction with probiotics. The most common prebiotics are nondigestible carbohydrates known as oligosaccharides. Prebiotics work in three ways: by feeding beneficial microbes; by aiding digestion and enhancing nutrient absorption; and by tricking pathogenic bacteria into attaching themselves to the oligosaccharides instead of to the intestinal wall. Because oligosaccharides are not digestible, they are excreted in chick poop, taking the bad bacteria along with them.

Prebiotics are beneficial with or without a probiotic. Prebiotics formulated for poultry are available from some poultry suppliers. Natural sources of prebiotics include chicory root, jicama, Jerusalem artichoke, asparagus, garlic, onions, dandelion greens, bananas, legumes (soybeans, lentils, kidney beans, chickpeas), wheat, barley, and raw oats.

The easy way to provide both prebiotics and probiotics — which is my choice, because I raise a lot of hatchlings each year — is to opt for a starter ration that contains both. Not many brands are formulated holistically, but that will surely change as poultry keepers become more knowledgeable about the nutritional needs of their birds.

SUITABLE FEEDERS

One of a hatchling's first instincts is to peck. Land-fowl hatchlings scratch as well as peck. Almost immediately on being placed in the brooder, they start looking for things to peck. If they are placed on newspaper, they will peck at the print. If they don't see anything to peck, they peck at their own toes.

To give hatchlings fresh out of the brooder something to peck, sprinkle a little starter on paper towels lining the brooder floor. Once they eat up all of the starter feed on the floor, they'll look around for more to peck and will find the feeders.

Chicks that arrive by mail may be weak and confused and need a little more help. For their first few days, feed them in a paper plate, a shallow tray, a shoe-box lid, or anything of similar chick height. Yes, they will waste some feed by pooping in it, but at this stage it's more important to get them started pecking for something to eat. Place feed no more than 24 inches (60 cm) from the heat source but not directly under it. When they start scratching out the feed, switch to a regular chick feeder, available from farm stores and through poultry-supply catalogs.

Like chick waterers, chick feeders come in several styles. An ideal feeder has these features:

- It prevents chicks from roosting over or scratching in feed
- It has a lip designed to prevent beaking or billing out
- It may be raised easily to the height of the birds' backs as they grow
- It is easy to clean

Round Feeders

Beyond the basic jar-and-base round feeder, which is suitable for only a couple of weeks at best, I have never found an ideal chick feeder. The basic feeder base is similar to a drinker base and comes in both galvanized steel and different colors of plastic. The base screws onto a feed-filled pint or quart (0.5 or 1 L) jar and has eight holes around the bottom through which hatchlings can peck. Because this style has a small footprint, it's ideal for use in a brooder where space is limited.

A similar-style eight-hole round feeder consists solely of closed base, with no provision for screwing onto a jar. It has the same footprint but holds considerably less feed, and its low profile invites chicks to roost on top, fouling the feed by pooping into the holes. On the other hand, the top pops off, making it easier to clean than the one-piece screw-on base. Round eight-hole feeders are good starter feeders for small brooders.

Tube Feeders

If the brooder is roomy enough, a tube feeder is ideal because it holds a lot of feed, so chicks are less likely to run out during the day; it minimizes feed wastage because chicks can't scratch in it and are less likely to bill out feed if the feeder is maintained at the proper height; and it is easy to hang so it may be raised to the height of the chicks' backs as they grow. Tube feeders come in various sizes, from small plastic ones to large metal ones. To determine how many chicks of age six weeks or under may be fed from one tube feeder, multiply the feeder's base diameter in inches (or centimeters) by 3.14 and divide by 2 (in metric divide by 5).

Talking Fowl

beaking out. A land-fowl habit of using the beak to scoop feed out of a feeder onto the floor

billing out. A waterfowl habit of using the bill to scoop feed out of a feeder onto the floor

withdrawal period. The minimum number of days that must pass from the time a bird stops receiving a drug until the drug residue remaining in its meat is reduced to an "acceptable level" — as established by the United States Food and Drug Administration, based on the drug manufacturer's recommendation — that is considered safe for human consumption

competitive exclusion. The process by which two species compete for available resources and the weaker one is displaced by the stronger one

FEEDER'S DIGEST

Jar lid

Tissue box

Eight-hole round feeder

Tube feeder

Jar and base

Trough Feeders

Trough feeders have a long, narrow footprint that takes up a lot of space in a confined brooder so are more commonly used in area brooders. They come in three basic styles: hole top, reel top, and grill top.

Hole-top trough feeders, like the basic round feeders, have separate holes through which individual chicks eat, which prevents squabbling and the wasting of feed. Hole-top troughs come in different lengths, typically 12, 18, and 24 inches (30, 45, 60 cm), having 15, 22, and 24 holes, respectively. Galvanized models have a sliding top that doesn't always slide easily — more than once I've cut my hand trying to slide one off or on — so like a lot of other people, I fill mine by pouring feed through the holes. Plastic hole-top troughs have a hinged top that flips open and snaps shut. Once the plastic hinges break, you're out of luck.

Hole-top troughs have the disadvantage that chicks will roost on top and poop through the holes, and sometimes a chick will climb into the trough through a hole and eat so much it can't get back out. Another disadvantage if you're raising broilers is that the chicks can't get their beaks through the holes after they reach about two weeks of age.

Reel-top trough feeders are made of plastic or galvanized steel and have an antiroosting reel across the length of the top that spins and is supposed to dump chicks that try to perch on it. A

Reel-top trough

Hole-top trough

single chick often has no trouble perching on the reel, although when other chicks try to join the first one, the reel turns and they dump each other off. In attempting to perch they can foul the feed. Like hole-top troughs, reel-top troughs come in different lengths. Some have adjustable legs so you can adjust the height as birds grow.

A **grill-top feeder** is basically an open trough with a grid on top to keep birds from walking or scratching in the feed. But chicks will try to scratch anyway, and as they get bigger the chance becomes more likely that they'll tip the trough and dump the feed. Screwing the trough to a length of 2×6 lumber stabilizes the feeder and prevents tipping.

The open design of both reel-top and grill-top troughs accommodates larger heads than hole-top feeders, so they are more suitable for larger breeds as well as for growing broilers that can't handily eat through the small holes.

With a hole-top feeder, allow one hole per bird. With an open trough, allow 1 linear inch (2.5 cm) per bird up to three weeks of age, 2 linear inches (5 cm) to six weeks of age, and 3 linear inches (7.5 cm) to twelve weeks. A trough that is not mounted against a wall and allowing birds to eat from both sides will accommodate twice as many birds.

Baby poultry of all kinds waste feed through beaking out or billing out — the habit of scratching out feed with their beaks or bills. To minimize waste fill the trough feeders only two-thirds full. An inwardly rolled lip discourages billing out, as does raising the feeder so it's always the same height as the birds' backs.

ADJUSTING TO FEEDING HABITS

Depending on how many birds you brood, and the size and style of their first feeder, as they grow you may need to switch to a larger feeder to make sure they can get their growing heads into the feeder and to ensure they don't run out of feed during the day. Whenever you change to a different feeder, leave the old one in place for a few days until you're sure all the chicks are eating from the new one.

Fill feeders in the morning, and let the chicks empty them before filling them again. Leaving feeders empty for long invites picking, but letting stale or dirty feed accumulate is unhealthful, so strike a happy balance. Clean and scrub feeders at least once a week.

Total Feed Intake to Six Weeks of Age*

	POUNDS PER BIRD	KILOGRAMS PER BIRD
Chick/keet	2.5	1
Poult	6	2.75
Broiler	9	4
Duckling	15	6.75
Gosling	20	9

*This intake table provides approximate amounts. The exact amount of feed needed to grow each bird to six weeks of age is influenced by numerous factors, including the bird's breed and strain, the brooding temperature, and the quality and nutrient balance of the feed.

Brooder Bedding

Hatchlings confined in a brooder accumulate a lot of poop. The more they grow, the more they poop. Keeping the brooder clean is about as pleasant a job as changing a baby's diapers, and just as necessary. Brooder bedding makes the task more manageable. But bedding that is suitable for growing birds is not the best bedding for the hatchlings' first few days in the brooder.

FIRST BEDDING

Until hatchlings learn what is edible and what is not, while pecking around they may fill up on bits of bedding, which jams up the works so edible feed can't get through, leading to starvation. For this reason, a lot of newbie chicken keepers are unreasonably afraid of using any kind of loose bedding. The trick is to make sure your birds are eating well before switching from their first bedding to loose bedding.

Hatchlings come out of the incubator looking for things to peck at. A solid surface with a little feed sprinkled over it will give them something edible to peck. The surface needs to be rough enough to prevent their little legs from slipping out from under them as they move around from the feeder to the drinker to their resting place near the heater. Newspaper or other smooth paper provides too little traction and can lead to leg injuries.

Choices

Plain white paper toweling may be unrolled in strips and used to line the floor of a small brooder. Some chicken keepers spread paper towels on top of shavings or other loose litter, but with a soft underbedding the chicks are apt to tear and wad up the paper. Instead, spread a thick layer of newsprint or other paper underneath the paper towels to help absorb moisture. As the paper toweling becomes soiled, simply add another layer on top. By the time the toweling gets messy faster than you can add a new top layer, the birds are big enough to get along without it. At that point either roll up all the paper and replace it with loose bedding, or spread loose bedding on top of the paper.

Chicken keepers who prefer to spend their money on soap and hot water, rather than on

Bedding for the Water Lovers

Waterfowl generate a lot more moisture than land fowl, creating a challenge in keeping them and their brooder clean and dry. Changing the bedding often enough to keep them healthily dry can get expensive.

Some waterfowl keepers prefer the bath-towel option, changing to fresh towels as often as necessary to keep the brooder environment healthful. Another option is to use puppy pee pads or human incontinence pads (also called bed underpads) to soak up moisture and odor. These pads are either washable and reusable or single use and disposable. The latter are available as scented or unscented pads. Loose bedding under the pads absorbs excess moisture. The most absorbent material for this purpose is pellet bedding, described on page 70. Sand is another option.

Some waterfowl keepers prefer to avoid messy bedding altogether by brooding their ducklings and goslings on hardware cloth with a water-collection pan, such as a plastic underbed storage tote, underneath the brooder. Excess water drips into the tote, which may be dumped and rinsed out as needed without having to remove the birds from the brooder. Most of the time the solid waste gets trampled on and pressed through the wire as well; if the brooder is overcrowded, however, the hardware cloth may clog up and need to be scraped off. A piece of plywood or other solid surface over a portion of the hardware cloth, away from the drinker, gives ducklings and goslings a comfortable resting spot.

paper towels, brood on old cloth towels or baby diapers. When a fresh towel is needed, the soiled one is shaken out, collected with other soiled towels, and laundered in the washing machine much as baby diapers are laundered. Avoid using big-loop terry towels, as hatchlings can get their sharp little toenails caught in the loops.

Nonadhesive, nonslip shelf liner is washable and therefore reusable. It is durable, yet soft and cushiony for hatchlings to rest and walk on. The rubbery nonslid surface is especially beneficial for chicks that have trouble with their little legs slipping out from under them.

Sold for gardening, interior decorating, and crafts, burlap makes a nice first brooder liner. Like nonslip shelf liner, it helps chicks get a good foothold to prevent their legs from sliding. Unless you can find a free or inexpensive source, the chief disadvantage to single-use burlap is the cost.

A piece of fine-mesh hardware cloth or an old window screen makes a good nonslip option for brooding chicks. In this case the wire is placed directly on top of a sheet of cardboard, several layers of newsprint or kraft paper, or over loose bedding. Cut the wire to fit the brooder bottom, and make sure the cut ends are turned under or taped to prevent injury to little feet.

In just a few days, usually less than a week, the brooded birds will have legs that are strong enough to let them be much more active, and they will know where to find edible feed. They will also generate greater quantities of poop, making the first bedding increasingly more difficult and time-consuming to maintain in a sanitary condition. At that point loose bedding becomes a better option. The little birds will peck and scratch in the loose bedding, and maybe carry bits of it around in their beaks or bills, but don't worry. Unless they are left with an empty feeder or can't find the feeder, typically they will not fill up on bedding.

Switch Bedding as Birds Grow

Paper towels (top) may be used as first bedding. After the birds are walking and eating well, switch to loose and more absorbent bedding, such as shredded paper (middle) or pine pellets (bottom).

LOOSE BEDDING FOR GROWING CHICKS

Once the birds start growing, a layer of good bedding will absorb their droppings to help keep them clean and dry, as well as insulate the brooder floor to retain warmth. Loose bedding also allows baby poultry to engage in natural activities such as scratching and pecking or dustbathing. Typical bedding depths are 1 to 2 inches (2.5–5 cm) in an enclosed brooder, 3 to 4 inches (7.5–10 cm) in an area brooder.

Ideal bedding is loose and fluffy but not dusty, absorbs moisture and droppings, has no objectionable odor, doesn't cake or mat, is nontoxic, and is easy for growing birds to walk on. Unfortunately, no one type of bedding is 100 percent perfect, but many suitable possibilities exist.

Wood Shavings

Dust-free kiln-dried pine shavings are popular as brooder bedding. They are suitable for area brooding and open-top brooders, but not for smaller and less well-ventilated enclosures. Pine contains phenols and other volatile compounds that can cause respiratory problems. But when the shavings have been well dried, most of the phenols have evaporated, and in an open area those that remain won't be a problem. You can easily identify shavings that have been properly dried because they don't have a strong pine odor. Cedar shavings smell stronger than pine because they contain more phenols, and therefore should not be used as bedding. Aspen and other hardwood shavings are not as absorbent as pine and are more expensive but lack phenols. Of course, never use shavings from treated lumber. Fresh sawdust should also be avoided, as it is dusty and can pack down and get soggy and musty. However, well-aged sawdust that is completely dry and free of mold can make acceptable bedding.

Pellet Bedding

Sold as stall bedding for horses and other animals, and as litter for cats, pine pellets are extremely absorbent, relatively odor free, less dusty than shavings, and last somewhat longer. The instructions on the bag suggest that the pellets should be moistened to fluff them up before use, but I use them straight from the bag and find that they absorb brooder moisture quickly enough on their own. Because they are so absorbent, they are one of the best bedding options for brooding waterfowl. Woodstove pellets are similar to bedding pellets, and some brands are sold as livestock bedding as well as stove fuel. Avoid stove pellets made of hardwood, which are not nearly as absorbent as pine pellets, and avoid brands that include an accelerant (a hazardous chemical additive that makes stove pellets easier to burn).

Shredded Paper

Newspaper or newsprint makes good bedding but must be freshened fairly often. Avoid using colored paper, particularly red or orange, because they make it more difficult to spot problems such as coccidiosis (a sign of which is bloody droppings).

shavings

pellets

paper

The smaller bits of paper shredded in a crosscut or microcut paper shredder are easier for baby birds to walk on than longer strip-cut paper, which can tangle around their legs, although the finely cut paper sticks to feathered feet.

Whole sheets of paper are not as absorbent as shredded paper, but if you have a ready and sufficient supply, paper sheets are acceptable for baby poultry that has grown past the first bedding stage. Several sheets of paper at the bottom of a brooder, beneath loose bedding or lining a litter pan, help make cleanup a little easier.

Grass Clippings

Well-dried grass clippings, from a lawn that hasn't been sprayed, make good bedding for land fowl that are not feather legged. Grass sticks to the feet of feather-legged breeds. And when used with waterfowl it gets soggy and matted.

When the weather calls for several warm, sunny days, I mow the grass in our orchard into windrows by driving the mower back and forth to blow the grass into narrow rows. The next morning I turn it with a pitchfork to dry the underside. Depending on the weather, it takes 2 to 3 days and one or two turnings to thoroughly dry the grass. Then I rake it up and pack it loosely in big clear-plastic bags that once held wood shavings. If the grass gets rained on before it's fully dry, or the bags bead with sweat inside, or the bagged grass smells the least bit moldy, I throw it on the compost heap. But when it's well dried and sweet smelling, grass makes acceptable brooder bedding. It mats rather

quickly, though, so it must be refreshed frequently by either replacing it or adding more on top.

Dry Leaves

Leaves, run over several times with a lawnmower to chop them up, make acceptable brooder bedding, provided they are fully dry and not the least bit moldy. Like dried grass, dried leaves tend to mat and must be refreshed frequently. I find, however, that mowing and drying leaves along with grass, as described above, makes fluffier bedding than either leaves or grass alone.

Chopped Straw

If you have access to clean straw, and have a way to chop it, chopped straw makes acceptable brooder bedding, but like leaves and grass it tends to mat and must be refreshed often. Regular straw is too long for baby birds to walk on without tripping, so use only chopped straw. If you don't have ready access to a straw chopper, you can chop small amounts at a time by putting the straw into a clean barrel or similar container and whipping it with a string trimmer. Clean (manure free) hay, especially hay that's been picked over by goats or other livestock, may be used similarly.

Peat Moss

Decomposed sphagnum moss harvested from bogs is sold as a mulch, soil amendment, and potting medium. It is soft and absorbent, is easy for baby birds to walk on, holds heat well to keep the brooder warm, and virtually eliminates brooding

grass   leaves   straw   peat

odors. Many people who brood chicks on peat feel it is superior to any other type of bedding. However, conservationists take issue with the fact that harvesting peat destroys wetlands that take eons to regrow.

Coir

Shredded coconut husk (coir, cocopeat, or coconut bedding) is widely used as a peat substitute. It is sold in pet shops and garden centers, and comes in compressed bricks or bales, or loose in bags. For brooder use the compressed form must by hydrated, broken apart, and dried. Improperly dried coir from any source contains a fungus that may be harmful to brooded birds. The safest coir to use in a brooder is the loose form sold as bedding for reptiles.

Vermiculite

Vermiculite is a natural mineral treated with heat so it expands, making it useful as a soil conditioner and growing medium. For more than occasionally brooding a few chicks, it can be pretty pricey. But folks who are willing to pay the price consider it ideal as brooder bedding because it is lightweight, sterile, and highly absorbent and won't catch fire should the brooder lamp fall into the bedding. A downside is the possibility that vermiculite may contain asbestos, since the two minerals are often found in the same mines. Residual dust in vermiculite is an indication that it contains asbestos, so use only refined vermiculite labeled as being dust free.

Sand

Clean mortar sand or sandbox play sand makes excellent brooder bedding. It is not as absorbent as other types of bedding, but it absorbs heat more readily and evaporates moisture more rapidly and therefore stays drier. Perhaps for the same reason, sand is more resistant to microbes than other types of bedding, keeping baby birds healthier. Soiled sand doesn't stick to feet like other types of bedding can. When it is kicked into a drinker, sand doesn't float and discourage drinking but rather sinks to the bottom so the water remains clean. Provided it is sifted periodically to remove chick poops, sand lasts much longer than other types of bedding, making it extremely economical. Like vermiculite, sand is fireproof, but because it retains warmth better than any other bedding, you have to be more careful about heater placement to avoid overheated chicks.

Pet Bedding

Pet-shop products such as CareFRESH (recycled, sanitized cellulose fiber), Eco-Bedding (recycled paper), and Cell-Sorb (recycled paper combined with gypsum) are designed to be super absorbent and odor free. They are expensive but last longer than all other options except pine pellets and sand.

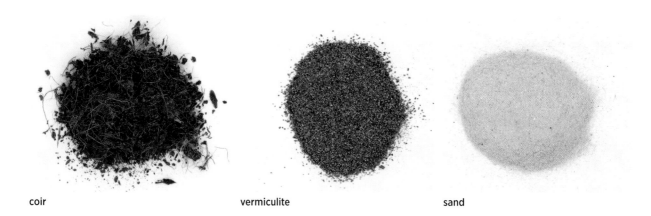

coir   vermiculite   sand

BEDDING MANAGEMENT

Managing bedding entails keeping it clean and dry. Additionally, loose bedding must remain fluffed up to retain its absorbency. Stir loose bedding daily to keep it from packing down, and add a little fresh bedding as often as necessary to keep it loose and absorbent. Remove and replace moist bedding around drinkers, since damp litter turns moldy and can cause brooder pneumonia. Remove and replace all the bedding as necessary to keep the brooder environment clean and dry.

In a small brooder, such as one that's storage tote size, if a thick layer of paper is placed on the bottom, before the bedding is added, most of the time the paper can be rolled up, litter and all, for removal. If the paper underneath gets wet, though, it may fall apart when you try to roll it up. In that case, a small brooder may be turned and dumped out (after removing the birds, of course). A larger brooder that's too heavy to lift may be cleaned out with a dustpan and a bucket or wheelbarrow. An area brooder or large brooder constructed at floor level may be emptied with a flat shovel, a coal shovel, or a snow shovel and a wheelbarrow or front-end loader. A cardboard box brooder is easiest of all to clean — just transfer the chicks to a fresh box and dispose of the old one.

As your birds become more active and their bedding gets deeper, some bedding will get kicked into the feeder and drinker. Raising the feeders and drinkers to a platform helps keep them clean. For this purpose I have used ceramic tiles, pieces of 2×8 lumber, bricks, paving stones, and hardware cloth platforms (described on page 58). Hanging the feeders and drinkers and raising them as the bedding deepens is an ideal way to keep them litter free.

If your brooder has a hardware cloth floor with a droppings pan underneath, the pan will be much easier to clean if you line it with paper, cardboard, or loose bedding. If you don't line the pan, you'll have the unpleasant task of periodically scraping off caked-on poop. When brooding waterfowl, a little bedding in the bottom of the pan will absorb some of the water to minimize sloshing when you move the pan for emptying and cleaning.

Used Bedding

How to dispose of used bedding can be an issue for urban and suburban chicken keepers. If you think you'll have a problem disposing of soiled brooder bedding, how will you deal with the greater quantities of used bedding generated by your poultry after they mature?

We gardeners who keep chickens love soiled bedding. It makes terrific compost for growing vegetables and flowers. If you aren't a gardener, surely you have a friend or neighbor or know a community or city garden volunteer who would love to have your used brooder bedding for their compost. They might even clean your brooder to get it.

The chief issue with adding used brooder bedding to a compost pile is that it contains a fair amount of starter ration, which can attract hungry rodents. Rather than dumping the bedding on top of the pile or burying it in one place inside the heap, mix the soiled bedding into the existing compost to make it less accessible and therefore less attractive to rodents.

4

WHAT TO EXPECT AS THEY GROW

Hatchlings don't stay cute and fluffy for long. They quickly start growing feathers, and before you know it they look like miniature versions of their mature selves.

Almost immediately from the time of hatch, baby birds display body conformation and behavior patterns characteristic of their species. At the same time, each bird develops a distinct personality and a unique tone of voice. If you have few enough birds that you can focus on individuals, you can distinguish it from the others — even if they all seem to look alike.

Early Chick Development

Hatchlings arrive in the world covered with down in place of feathers on most parts of their bodies, with the exception of tiny early feathers on their wings and tails. Down provides insulation to keep the babies warm. It also provides camouflage that lets them blend in with the environment — baby poultry are not the same bold colors they will be as adults.

Down grows from follicles, similar to those from which human hair grows. Mature poultry also have down on some parts of their bodies, but that down grows from follicles that specialize in the sole production of down. Some of a hatchling's down grows from the same follicles that will eventually sprout feathers. Baby down, in fact, consists of the tips of feathers. As the feathers grow longer, the down breaks away.

Newly emerging feathers are pointed, like a pin, so they are typically called pinfeathers. Blood vessels extend into the feather shaft to nourish these developing feathers, which are therefore sometimes called blood feathers. Each dark, blood-filled feather is tightly packed into a keratin sheath — a thin tubelike structure that eventually splits open and falls off, allowing the feather to unfurl.

Once a feather is fully formed, the blood vessels die back and the feather shaft hardens to better protect the bird's skin. The feather follicles are linked by a network of tiny muscles that allow a bird to raise and lower its feathers; for instance, to trap warm air by puffing out the feathers in cold weather.

Growing Pinfeathers: The Awkward Age

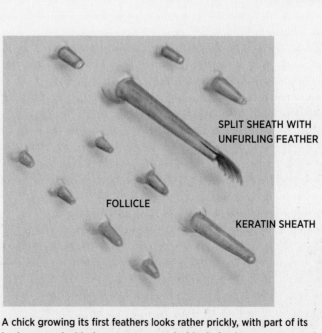

SPLIT SHEATH WITH UNFURLING FEATHER

FOLLICLE

KERATIN SHEATH

A chick growing its first feathers looks rather prickly, with part of its body covered with down, part covered with pin feathers, part covered with maturing feathers, and part bearing no feathers at all.

FEATHER TRACTS

Unlike a mammal's fur, feathers do not grow over the entire body surface of a chicken or other type of poultry. Instead, the feathers grow in small, symmetrical tracts and fan out to cover the body. The feather tracts (technically, pterylae) are separated by featherless areas (apteria) in which some down may grow, even in mature poultry.

The featherless areas facilitate cooling when a bird holds out its wings in hot weather to expose bare skin underneath. Otherwise, no one is exactly sure why poultry and other birds grow feathers in distinct tracts. One theory is that feather tracts minimize the total number of feathers, thus reducing the bird's overall weight to make it better able to fly. This theory is supported by the fact that flightless birds — penguins, along with ostriches and their relatives — do not have feather tracts but grow feathers all over their bodies.

Many types of barnyard poultry no longer fly well or at all, but they all still grow feathers in tracts. The specific shape and distribution of tracts differs for different bird families. A chicken has 10 distinct feather tracts — head, neck, shoulder, wings, breast, back, abdomen, rump, thigh, and legs. Just as no one is exactly sure why feathers grow in tracts, no one has yet determined why the arrangement of tracts differs from one family of birds to another.

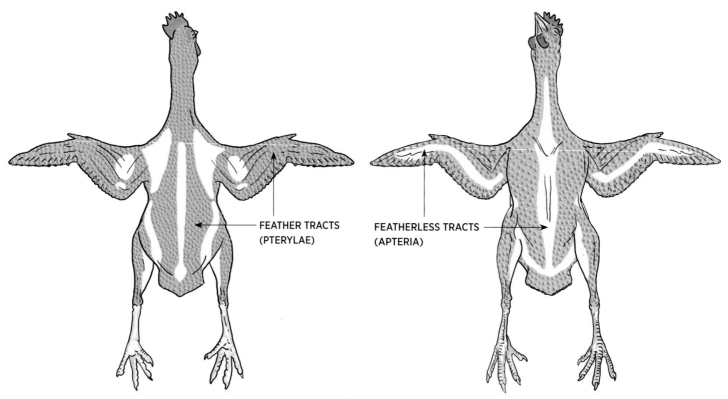

FEATHER TRACTS
(PTERYLAE)

FEATHERLESS TRACTS
(APTERIA)

Broilers Grow at a Different Rate

The growth rate of broiler strains is a bit different from other types of chickens. They put much less energy into feather growth and much more energy into muscling up. As a result they get big considerably faster than other strains and have a much sparser layer of feathers compared to other birds of the same age — their feather growth simply can't keep up with their rapid increase in size. Fewer feathers means easier plucking when the time comes for the harvest. But even though broilers are short on feathers, their heavier bodies cause them to readily overheat.

First Feathering

All poultry develop their first coat of feathers in about the same order. The wing feathers are important for mobility, so by the end of the first week, baby birds have distinguishable little wings. The tail helps with balance, so by the end of the second week, you'll see a pert little tail. Body feathers are important for warmth, so they gear up during the third week and fill out in the fourth week. By the end of the fifth week most of the feathers are in, and the birds look like little chickens, guineas, turkeys, ducks, or geese.

Just about the time they are fully feathered, they molt — some of their feathers drop out sequentially, to be replaced by new feathers. Before the birds reach maturity, they go through another partial molt. Some breeds, especially the fancy show models, don't develop their characteristic feather patterns until after this stage.

Species, breed, feed, and environmental temperature influence the exact age at which they go through these various stages. The cooler the temperature, the more rapidly feathers grow. Keeping the brooder temperature on the cool side of comfortable encourages rapid feather growth; failing to reduce the temperature as the birds grow delays full feathering.

Talking Fowl

feather tract. A specific area of the skin in which feathers grow

full molt. The renewal of feathers in all the feather tracts

keratin. Fibrous protein that forms the structural basis for feathers and claws

molt. The periodic shedding and renewal of plumage, controlled by hormones and regulated by daily exposure to light

partial molt. The renewal of feathers in some of the feather tracts

Typical Chick Feathering Timetable

WEEK	FEATHER GROWTH
1	Wing
2	Tail, shoulder
3	Breast, back
4	Body
5	Head

Age at Full Feathering

SPECIES	AVERAGE AGE
Chick	5–6 weeks
Keet	6–8 weeks
Poult	6–8 weeks
Duckling	6–8 weeks
Gosling	12–15 weeks

RATE OF GROWTH

Some species, breeds, and strains grow much more rapidly than others. Strains of all species that have been bred for meat production grow the fastest of all. Old-time breeds, especially the ones that mature to the largest sizes, grow at the slowest rate.

The accompanying series of photographs shows typical growth and feathering rates for Rhode Island Red pullets, Rhode Island Red cockerels, and Cornish-Rock broiler cockerels. The broilers were raised in the same brooder environment and on the same rations as the Rhode Island Reds, with no attempt to influence their growth rate. Each week a few representatives were taken from each group for the photo op. Selecting a single individual from each group to photograph proved impractical, because no bird was consistently cooperative from week to week. Similarly, each bird gained weight at a slightly different rate, so the weekly weights represent group averages. These photos indicate what you can expect as your chicks grow.

MY, HOW FAST THEY GROW!

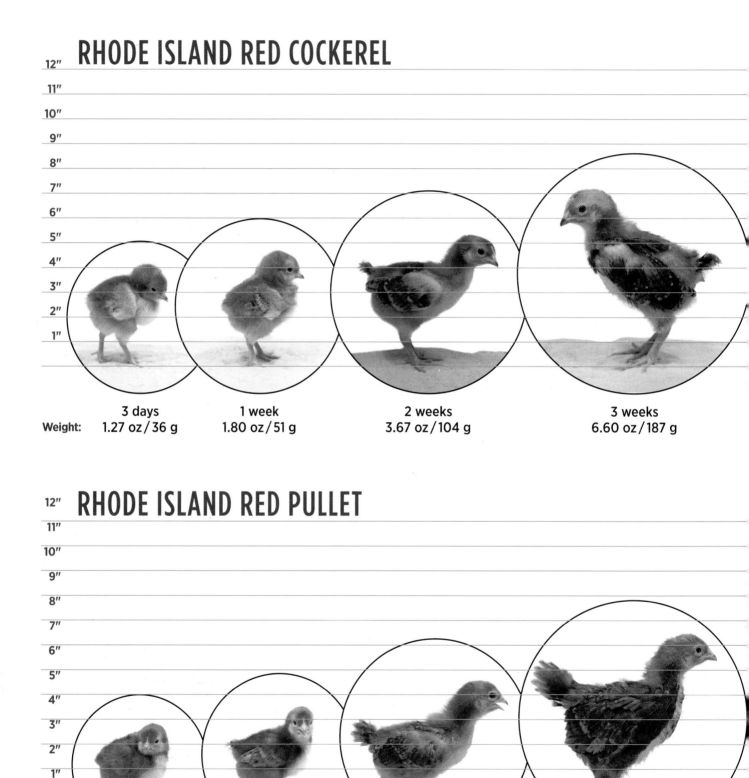

RHODE ISLAND RED COCKEREL

12"
11"
10"
9"
8"
7"
6"
5"
4"
3"
2"
1"

	3 days	1 week	2 weeks	3 weeks
Weight:	1.27 oz / 36 g	1.80 oz / 51 g	3.67 oz / 104 g	6.60 oz / 187 g

RHODE ISLAND RED PULLET

12"
11"
10"
9"
8"
7"
6"
5"
4"
3"
2"
1"

3 days	1 week	2 weeks	3 weeks
1.23 oz / 35 g	1.73 oz / 49 g	3.63 oz / 103 g	6.77 oz / 192 g

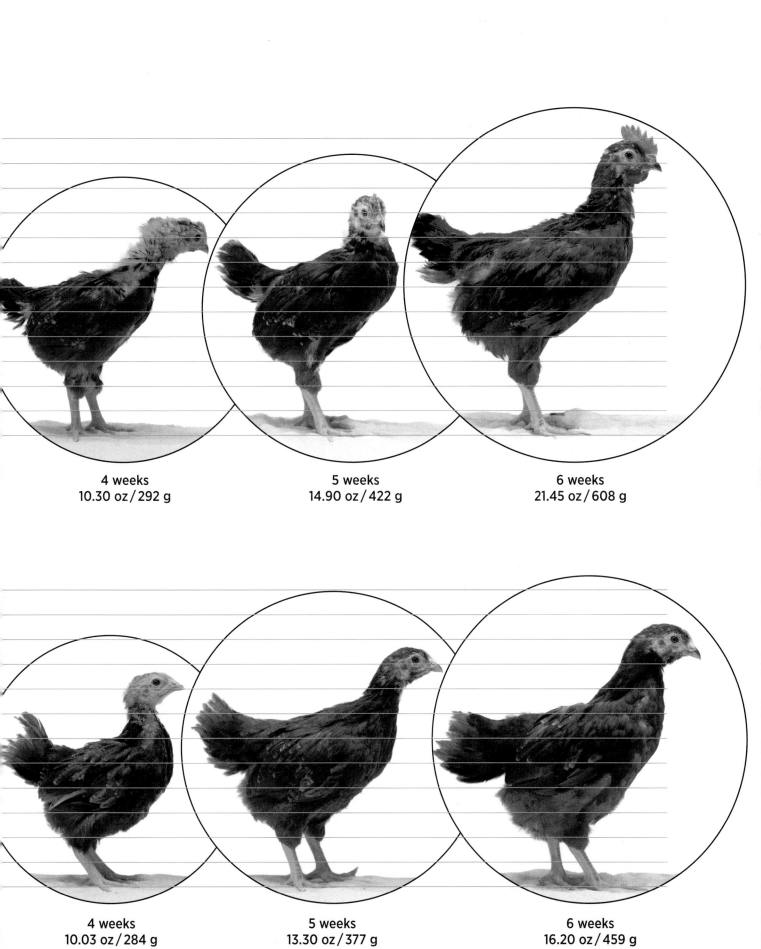

4 weeks
10.30 oz / 292 g

5 weeks
14.90 oz / 422 g

6 weeks
21.45 oz / 608 g

4 weeks
10.03 oz / 284 g

5 weeks
13.30 oz / 377 g

6 weeks
16.20 oz / 459 g

CORNISH-CROSS BROILER

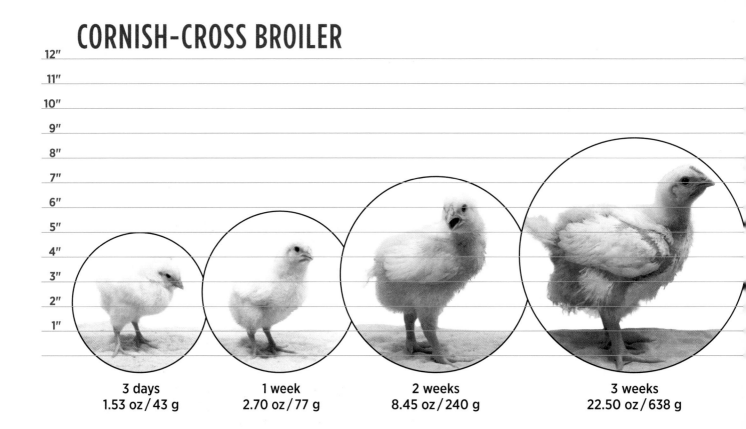

3 days	**1 week**	**2 weeks**	**3 weeks**
1.53 oz / 43 g	2.70 oz / 77 g	8.45 oz / 240 g	22.50 oz / 638 g

Taming Young Birds (and Children!)

If you want your birds to be pets, you'll need to spend time with them so they get used to being around people. But don't handle them during the first day, as they need to rest and to get oriented to the brooder environment.

If young children are involved, make sure they know not to handle the birds roughly, squeeze them, or drop them. The best way for a really small child to get acquainted with baby poultry is to have the child sit on the floor with a towel across his or her lap (to catch stray poops) and place a baby bird on the towel where the child can touch it without actually holding it. Children, especially young ones who can't remember to keep their hands out of their mouths, should always wash their hands after handling the babies or helping maintain the brooder.

The more time you spend with your baby poultry, the friendlier they will become. That can be a good thing later on, when you need to handle them to check their health or to collect eggs from beneath the hens. With goslings, however, too much handling early on is not such a good thing, as they can get decidedly nasty when they mature; they need to learn to have a healthy respect for humans.

With all brooded poultry, take care not to overdo the handling. Hatchlings are, after all, babies that tire easily. Let them spend most of their time like any babies — eating or sleeping. As they grow and feather out, continue handling them with care. Young poultry need to conserve most of their energy for growing muscle, bone, and feathers.

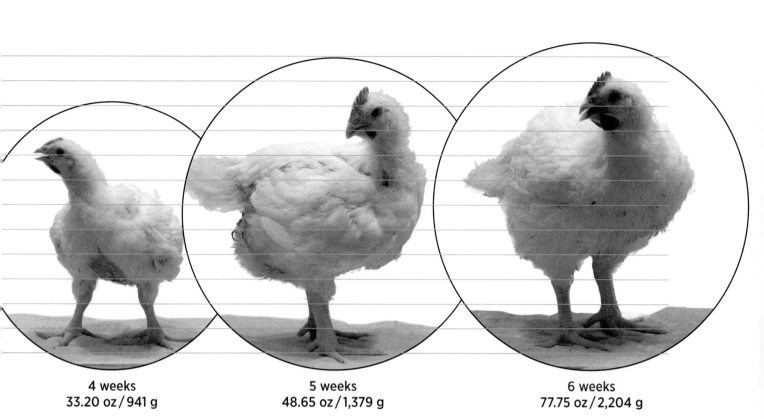

4 weeks	5 weeks	6 weeks
33.20 oz / 941 g	48.65 oz / 1,379 g	77.75 oz / 2,204 g

Day Trips and Camping Out

As your little birds feather out, they will enjoy time outdoors during warm weather. Unless they are supervised the entire time they are outside, however, they will need protection from passing dogs, cats, hawks, and anything else that might take a fancy to a quick snack. For complete protection they must be kept in a secure enclosure, such as a wire-bottom cage or a doghouselike structure with an attached enclosed run.

Provide water during their outing, and if they will be out most of the day, also bring along their feeder. For protection from the hot sun, they'll need some form of shade, which could be as simple as a towel draped along part of the cage or run. They will also need protection from breezes; on downright gusty days keep them indoors. If the birds huddle or act uncomfortable in any way — appearing to be too hot or too cold — bring them back to their brooder and try again another day.

When daytime temperatures remain above 65°F (18°C) and nights are not overly chilly, broilers, ducks, and guineas may be moved outdoors permanently when they are about four weeks old; chickens, geese, and turkeys when they are about six weeks old. At this age they are too young to be turned out the graze, but may be kept in a coop with a run that offers protection from hot sun, wind, and rain and provides a sheltered place in which to spend the night. In sufficient numbers, they may keep each other warm on cooler nights.

For the first few days, check on them at night to make sure they are okay. If the nighttime coop temperatures dip close to freezing, or if the weather turns stormy, gather them up and bring them back into protective custody. Once they are fully feathered, they can remain in unheated, outdoor housing.

Separating the Sexes

If you are raising straight-run poultry, at some point you might consider separating the males from the females. Although the natural gender ratio at hatch is approximately 50/50, in the average backyard flock fewer males are needed than females. You don't need a male at all, unless you plan to hatch eggs in the future. The male does not *cause* a hen to lay eggs but merely fertilizes the eggs she does lay so they will hatch.

All species raised in small backyard numbers may be kept in pairs, although trios (two hens and one male) are more economical in the long run, and therefore more typical. For a backyard flock, the table (below) shows the optimum number of females per male of each species. If there is an excessive number of males in a flock, they will harass the females and be overly aggressive toward one another.

SEXING CHICKENS

The gender of chicks may be determined by a number of different methods, depending on the breed and variety, the skill of the observer, and the age of the chicks. Vent sexing is done soon after hatch. Color sexing and feather sexing may also be done at hatch but remain apparent for some time after hatch. For many breeds, visible differences between the sexes (sexual dimorphism) begin to develop as the birds grow.

Vent Sexing

Vent sexing, also called cloacal sexing, is a method of determining a hatchling's gender by examining minor differences in the tiny cloaca just inside a chick's vent. Accuracy requires a great deal of training, skill, and keen observation.

It is usually done at hatcheries by trained sexers who sort the cockerels from the pullets for shipment to customers who place orders other than straight run. Vent-sexed bantams are not often available because bantams are so tiny and delicate they easily may be injured in the process. A good vent sexer has an accuracy rate of 95 to 98 percent.

Color Sexing

Color sexing takes advantage of sex-link genes that control down color. Color sexing within a single breed or variety is called autosexing. For instance, barred and cuckoo chicken varieties all look pretty much alike — black with white spots on the head — but the cockerels are a little lighter in color, and the spots on their heads tend to be irregular and elongated or scattered, while the pullets are somewhat darker and their spots are more round and compact. Depending on the breed, autosexing has an accuracy of 80 to 95 percent. Breeds, such as the California Gray, that have been deliberately developed for this feature may be sexed at near 100 percent accuracy.

Hybrid chicks that may be sexed according to color are called sex-links. The most common purpose for breeding sex-links is to obtain near-industrial-strength laying hens that have characteristics not found in standard commercial-strain White Leghorn layers. Crossing a California Gray cock with a White Leghorn hen, for instance, produces the sex-link California White. The cuckoo

Best Mating Ratios

FOR EACH MALE OF THIS BREED OR SPECIES	OPTIMAL NUMBER OF HENS
Chicken, light breed	12
Chicken, heavy breed	8
Guinea	5
Turkey	5
Duck, light breed	7
Duck, midweight breed	6
Duck, heavy breed	5
Muscovy	5
Goose, light breed	5
Goose, heavy breed	3

California Gray puts head spots only on the resulting California White pullets. California White hens are more cold hardy than Leghorns, but like Leghorns they lay eggs with white shells.

Except for California Whites, all other commercial-quality sex-links lay eggs with brown shells and are known as either black sex-links or red sex-links. Black sex-links are a cross between a red cock (usually Rhode Island Red or New Hampshire) and a barred Plymouth Rock hen. At hatch all the chicks are black, but only a cockerel has a white spot on its head. Trade names for black sex-links include Black Star and Black Rock.

Red sex-links are produced by crossing a red cock with a silver-based (white) hen such as a Delaware or a white Plymouth Rock. The chicks may be sexed easily at hatch because the males are mostly white, while the females are either buff or red. Trade names for red sex-links include Cinnamon Queen and Golden Comet. Sex-links can be sexed with 99 to 100 percent accuracy.

Autosex Breeds

BREED/VARIETY	MALE	FEMALE
Chickens		
Barred and cuckoo breeds/ varieties	Paler; missing, large, irregular, elongated or scattered head spots	Darker down and legs; small, compact, narrow head spots; distinctly yellow toes
Black-breasted red, silver duckwing breeds/varieties	Lighter dorsal stripes ending in dot at crown	Dark dorsal stripes along back to crown
Buckeye; New Hampshire; Orpington, buff	White or cream-colored spot on upper wing	Dark spot on head or dorsal stripes along back
Rhode Island Red	White or cream spot on wing and belly	Darker red down color
Leghorn, light-brown and silver; Cornish, dark; Welsumer	Lighter dorsal stripes, sometimes ending in dot at crown	Dark dorsal stripes along back to crown
Norwegian Jaerhon, dark	Brown; larger head spot	Brown; smaller head spot
Norwegian Jaerhon, light	All yellow	Yellow; brown stripe along back
Penedesenca, crele	Gray	Brown
Geese		
Embden; Roman; Sebastopol	Pale gray and yellow	Darker gray
Pilgrim	Yellow	Gray
Shetland	Yellow	Gray and yellow

Talking Fowl

autosex. A straightbred variety or breed that displays clearly distinct sex-linked color characteristics by which cockerels may be distinguished from pullets at the time of hatch

barred. A color pattern of feathers consisting of alternate crosswise stripes of two distinct colors

cloaca. The chamber just inside the vent where the digestive, urinary, and reproductive tracts come together

cuckoo. Coarse, irregular barring

dorsal stripes, also called *chipmunk stripes*; stripes that run along the back, usually on both sides of the backbone

hen feathered. A cock's plumage that is nearly identical in color and markings to a hen's of the same breed and variety

primary feathers. The 10 largest feathers of the wing

sexing. Separating the cockerels from the pullets

sex link. A hybrid that has been crossbred to take advantage of sex-linked genes so its gender may be determined at hatch by physical appearance (color sexing or feather sexing), rather than by a microscopic examination of its sex organs (vent sexing)

sex-linked genes. Genes that are carried on the pair of chromosomes that determine a bird's gender

vent. The anal opening

Feather Sexing

Recognizing growth rate differences in the primary feathers of hybrid chicks is the key to feather sexing. Crossing a slow-feathering hen (such as a Rhode Island Red) with a rapid-feathering cock (such as a White Leghorn) yields slow-feathering cockerels and rapid-feathering pullets. The chicks are all the same color but may be sexed by the appearance of well-developed wing feathers on the pullets at the time of hatch. Sexing by this method has an accuracy rate of about 99 percent.

Feather sexing is common in the broiler industry — where white-feathered birds are preferred — so the slow-growing pullets may be raised apart from their faster-growing brothers. The difference in the growth of the primary feathers is seen only between one and three days of age, after which the cockerels' wing feathers catch up with the pullets' and they all look alike.

Sexual Dimorphism

Observable differences in the physical features and behavior patterns between males and females is called sexual dimorphism (di = two; morph = forms). At three weeks of age or older, depending on their breed, chicks start developing reddened combs and wattles. The cockerels' combs and wattles will become larger and more brightly colored than the pullets'. Unless they're Sebrights or Campines, which are hen-feathered breeds (the males and females have nearly identical plumage), the males will soon develop pointed back and saddle feathers and long tails, in contrast to the more rounded back and saddle feathers and shorter tails of hens.

At about the same time as the cockerels start developing combs and wattles, they make their first squeaky attempts at crowing. When crowing starts in earnest, peck-order fighting will get serious and sexual activity will start. It's time to separate the cockerels from the pullets, or at least pare down the number of cockerels to a reasonable ratio for the number of pullets. Since many fewer cockerels than pullets are needed, nearly everyone who keeps chickens has a surplus. The traditional way to deal with them is to put them into the freezer. The accuracy rate of sexing by this method increases to nearly 100 percent as the birds mature.

Vive la Différence!

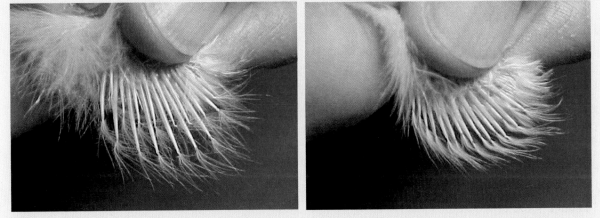

Feather sexing is based on the growth rate of the primary feathers; a pullet's feathers (left) develop more rapidly than a cockerel's (right).

Sex link hybrid chicks may be sexed according to their down color. Red Star is one of many examples: the cockerels (left) are white, the pullets are red.

These month-old Old English Game demonstrate sexual dimorphism: the cockerel (left) has a larger, redder comb and wattles, pointy back and saddle feathers, and a longer tail than the pullet.

SEXING POULTS AND KEETS

Poults and keets are normally sold as straight run, with the exception of broad-breasted meat turkeys, which some hatcheries will vent sex. Otherwise, males and females of both species, within the same breed or variety, look pretty much alike until they are nearly mature.

As poults grow you can take an educated guess based on a few criteria. Among poults of the same breed, the little hens are somewhat daintier, while the little toms generally weigh slightly more and have slightly larger feet, slightly thicker shanks, and somewhat less feathering on their heads. A nearly surefire way to tell the toms from the hens is to watch for strutting, which can start at any age. Some toms will strut nearly as soon as they start walking. They'll drop their wings, tuck in their heads, raise their stubby tail feathers, and show off just like a mature tom. Hens occasionally strut, too, but much more briefly and usually not until they are of breeding age.

Keets are even more difficult to sex. Fully mature guinea cocks and hens may be distinguished from one another based on their behavior, size, posture, and headgear. But these differences don't become apparent until they are nearly full grown. In Buff and Buff Dundotte varieties, the females are somewhat darker than males; to a lesser extent the same is true of brown and chocolate varieties. Otherwise, the only surefire way to tell young ones apart is by the call made by females, starting at about six weeks of age.

Their call is often said to sound like "buckwheat." Although we commonly say the word buckwheat with the accent on the first syllable — *buck*-wheat — female guineas put the accent on the second syllable — buck-*wheat* — sounding more like "good *luck*" or "come *back*." Male guineas never make this two-syllable call, although if you raise a group of keets, it's nearly impossible to identify the ones doing the calling. However, at about the same time the females start calling, the males' heads take on a subtle blockier appearance compared to the females' more refined heads.

A young tom drops his wings, tucks in his head, raises his stubby tail feathers, and struts just like his daddy.

SEXING DUCKLINGS AND GOSLINGS

Male and female waterfowl of the same breed and variety are difficult to tell apart until they are several weeks old. Except for a few autosex varieties of geese, the only way to determine the gender of waterfowl hatchlings is by vent sexing, which is not nearly as difficult as with land fowl.

As ducklings grow, darker bills may distinguish the hens of nonwhite varieties; the hens of Mallard and similarly colored varieties have brown and orange bills compared to the drakes' greenish bills. As they feather out, the head and back

feathers of Mallard-colored hens are not as dark as those of the drakes. While the call of the female matures to a loud quack, the male's call becomes a hoarse whisper. As the drakes lose their voices they develop conspicuous drake feathers that curl up and forward at the top of the tail. Among Muscovies the drakes are much larger than the hens.

Distinguishing a male from a female gosling is tricky, except for the autosex breeds — Pilgrims and Shetlands, and to a limited extent Embdens,

VENT SEXING WATERFOWL

Vent sexing is the only reliable way to sex non-autosex waterfowl hatchlings, but it requires lots of practice, as well as extreme care to avoid injuring delicate parts. Having an experienced person show you how to do it is the best way to learn. Barring that, practicing on ducklings or goslings that are a couple of weeks old lets you learn with less chance of injuring the older birds and a better chance of seeing what you're looking for.

When sexing, you're trying to detect the presence or absence of a penis. Because the organ is so tiny, you need a strong light and good eyesight (or a good desktop magnifying glass, some of which come with a built-in light). Understandably, the little fellas are reluctant to show you theirs, so just because one doesn't pop out does not automatically mean the bird is a female. To know for sure you need to peer into the everted cloaca to ascertain whether or not a penis is in there. Proceed with the following steps:

1. Gently pick up one bird and turn it upside down so you can see the vent easily.

2. Bend the tail toward the bird's back, out of the way. Don't be surprised if a little bit of poop squirts out.

3. With your thumbs smooth the down away from the vent so you can get a better view.

4. Use your thumbs and one index finger to spread the vent to expose the cloaca; do not squeeze the vent as if it were a pimple.

5. If it's a male, a translucent wormlike penis will corkscrew out; a female has, instead, a rosette of pink tissue known as the genital eminence.

6. If you can't tell for sure after about 30 seconds of manipulation, release the bird and try again later.

The average gender ratio is about 50/50. Newbies learning to vent sex commonly find far more females than males, which means some of those identified as females are actually shy males. An excellent vent sexing video showing how to hold a bird and distinguish between males and females may be found online at www.metzerfarms.com/SexingVideo.cfm.

To sex a duckling, use two thumbs and an index finger to stretch the vent (left). Do not squeeze the vent.

A male has a little corkscrew penis.

A female has a genital eminence.

Romans, and Sebastopols. Among Pilgrims and Shetlands, the males and females maintain two different plumage colors throughout their lives — the ganders are yellow when young and feather out white, while the hens are gray or gray and white. Embdens, Romans, and Sebastopols may be sexed with a fair degree of accuracy for only about their first three days of life — when the hens are a darker gray than the ganders — after which both genders are pretty much the same color.

Among African and Chinese geese of about six weeks of age, or sometimes earlier, the head knob of the male becomes noticeably larger than that of the female and remains the distinguishing feature by which the genders may be told apart into maturity. For the breeds that are neither autosex nor knobbed, male and female geese are difficult to tell apart until you gain enough experience to recognize subtle differences in voice and body posture among the goslings.

Instinctive Behaviors

Hatchlings and growing poultry display certain instinctive behavior patterns that some novice poultry keepers find odd and disconcerting.

Understanding what is normal behavior allays concerns and lets you more readily recognize abnormal behavior that may require intervention.

BABY CHICKEN SOUNDS

A bird peeps even before it hatches from the egg, and shortly after hatching, it makes a number of different sounds. Although each species has its own distinct set of calls, they are basically similar in function. Happy sounds tend to swing upward in pitch, while the unhappy sounds descend in pitch. The sounds made by young chickens have been studied more than those made by other poultry and have been identified as follows:

Pleasure peep is a soft irregular sound chicks use to maintain contact with each other and, when raised by a hen, to maintain contact with their mother.

Pleasure trill is the soft, rapidly repeated sound chicks make when they've found food or are nestling into a warm, safe place to sleep.

Distress peep is a loud, sharp series of sounds chicks make when they're cold or hungry.

Panic peep is a loud, penetrating peep of a chick that's scared or lost. It is similar to the distress peep but louder and more insistent.

Fear trill is the sharp, rapidly repeated sound made by a chick that sees something strange or potentially threatening, such as a hand reaching to grab it.

Startled peep is the sharp, surprised cry of a chick that's been grabbed abruptly.

Talk to Your Hatchlings

Talking to baby birds in your brooder helps keep them calm by letting them know you are coming before you suddenly appear. Make sounds that are low pitched, brief, soft, and repetitive to attract, calm, and comfort them. Avoid sudden or high-pitched, long, and loud sounds, which baby poultry find frightening.

Poultry begin establishing the peck order at an early age.

PECKING ORDER

The peck order is a dominance hierarchy maintained by birds of higher status pecking those of lower status. By governing a flock's social organization, such that each bird knows its place within the hierarchy, the peck order minimizes tension and stress. In a flock containing both genders, the peck order involves a complex hierarchy on three levels: among all the males, among all the females, and between the males and the females. A new bird added after the pecking order has been established must work its way up the ladder but doesn't necessarily start at the bottom.

Baby poultry begin establishing the pecking order soon after hatch, but the serious jostling doesn't ramp up until they are several weeks old and starting to mature sexually. Peck-order fighting among brooded poultry is usually not serious enough to cause significant injury. Most of the time two little cockerels face off or two little drakes shove each other around, until one loses interest and turns away. Separating the genders, or at least reducing the number of excess males in a straight-run group, minimizes such skirmishes.

IMPRINTING

Baby poultry instinctively follow their natural mother and may be induced to follow a substitute foster mom. Ducklings and goslings, in particular, develop a strong attachment to whomever they first see upon hatching. Young poultry that imprint on a human keeper will follow the human, even after they mature. If your goal in keeping poultry is to have backyard pets, imprinting gets them to bond with you from an early age and eliminates the need to spend time taming them later.

Land fowl take much more time to imprint than waterfowl. Ducklings imprint rather easily, and goslings imprint almost too easily. I once raised a brood of Embden goslings that were convinced, through no deliberate attempt on my part, that I was their mother. When time came to release them into our pond, they didn't want to go. They kept following me back to the brooder house. Even after I managed to sneak away, for several weeks thereafter whenever they heard my voice they'd coming running in my direction.

Imprinted goslings may be cute when they're young, but when they mature they can be frustrating and even downright dangerous. The apparent purpose of imprinting is to allow a bird to recognize its own species by forming an attachment to its mother and, later, to find a suitable mate. At some point it becomes more appropriate for geese to be geese than to continue believing they are humans. But even if you are able to convince them to find mates among themselves, when nesting starts the ganders may challenge you in attempting to defend their mates. Believe me, you do not want to be attacked by an aggressive gander. Better to not get too cozy with your baby goslings at the outset, but let them learn to trust you while recognizing you as being higher in the peck order.

Similar issues may arise with Muscovy ducklings and with turkey poults but rarely with other ducks or land fowl. The bigger species defend themselves by being aggressive; the smaller ones by hiding. Taking advantage of imprinting to tame ducklings, chicks, or keets therefore makes more sense. The more time you spend with them, the more strongly they will bond to you.

BATHING

Land fowl enjoy taking dust baths, which help clean their feathers, minimize external parasites, and give the birds something pleasurable to do. They squat or roll in dry soil or bedding and use their feet to kick it over their bodies. When they are done, they get up and shake themselves off; the dust flies, and so do dead skin cells and other debris that accumulates on their bodies.

Given suitable bedding, chicks will enjoy dust baths when they're only a few days old. Typically, when the first bedding of land-fowl chicks is changed to loose bedding, the delighted birds will engage in an orgy of dustbathing.

If the dust wallow is in a warm sunny place — such as a patch of sunlight streaming through a window — sometimes a bird will cover itself with dust, then stretch out and just lie there for a time, soaking up the sun. Coming across one or more birds thus languishing can be a bit of a shock, as they appear to be quite dead until they jump up at your approach.

Waterfowl, as you might expect, prefer to bathe in water. Raised naturally, ducklings and goslings will happily swim with their mothers. The matriarch controls how long the babies stay in the water and when they must come out. Left on their own, however, ducklings and goslings tend to play in the water long after they should come out, becoming soaked and chilled, and sometimes even drowning. Because their waterproofing oil gland doesn't develop until they are a few weeks old, wet down may stick together and waterlog. Brooded waterfowl should never have unsupervised access to water deep enough to swim in, which includes drinkers of such a design that the birds can climb in.

Swimming is so instinctive that baby waterfowl lacking access to water deep enough to swim in will dip their bills into the drinking water and make bathing motions. Occasional brief supervised baths in shallow, clean water — in a sink, dishpan, bathtub, or kiddy pool — are just as much fun for them as for you to watch. Just be sure the ambient temperature is warm enough to prevent a chill, and return them to the warm brooder before they get soaked through.

Baby waterfowl enjoy brief supervised bathing sessions in shallow, clean water.

FEEDING RITUALS

Eating is one of the two main activities of brooded birds, the other being sleeping. Each species has its own ritualistic behavior patterns related to feeding, and baby birds engage in those characteristic patterns nearly from the start.

Ground scratching stirs up the soil to bring up seeds, insects, worms, and bits of grit. Baby land fowl that get their first starter ration served in a paper plate or on a paper towel will instinctively scratch in the feed. Initially, that's okay, because the most important thing is to interest them in eating. But in scratching they scatter feed out of the tray, walk in it, and poop in it. And that's not okay. So as soon as they are eating well, switch to a regular chick feeder they can get their heads into but not their feet.

Food running is characteristic of most poultry species, supposedly having the purpose of breaking up a large edible item into pieces small enough to swallow. A chick that gets hold of something tasty that's too big to swallow may pick it up in its beak and start running. Other chicks give chase, try to grab the tidbit, and engage in a tug-of-war until pieces break off (or one of the other chicks gets hold of the whole thing and runs with it). A chick that finds a worm, for instance, may grab it and start running, thus attracting others to chase after. Toss a cooked spaghetti noodle into the brooder and be prepared for a jolly good round of food running.

Food running does not always involve edible items. A chick may pick up a piece of bedding in its beak and carry it around or attract attention by running with it. That doesn't imply the chicks are eating bedding. They are merely engaging in instinctive behavior that we humans call food running.

Beaking out or billing out is the practice of using the beak or bill to scoop feed from a feeder onto the ground, and it is common to both land fowl and waterfowl. Feed on the ground is usually wasted because it gets trampled into the bedding and covered with droppings. No one knows exactly why birds do this. Perhaps they're hoping to find something more interesting to eat, or maybe they're just bored.

Two things encourage the practice — improper feeder design and inappropriate feeder height. A feeder with a deep trough and a rolled-in or bent-in edge reduces wastage, but unfortunately, most feeders lack these features. The primary way to discourage billing out is to keep feeders at the height of the birds' backs so they don't have to bend over far to eat, which means raising feeders as the birds grow. A hanging tube feeder is easiest to adjust. Some trough feeders have adjustable legs. If you use open-top trough feeders, a third way to reduce feed wastage is to never fill the troughs more than two-thirds full, which keeps the feed level low enough to minimize scooping.

PERCHING

By the time light breeds of land fowl reach four weeks of age and heavy breeds about six weeks, they're ready to roost on low perches. Don't provide roosts for broilers or they'll develop breast blisters and crooked keels. And of course, ducklings and goslings (with the exception of Muscovy ducks) don't perch.

For others nighttime roosting is a natural and healthy habit, and hopping on and off perches during the day is good entertainment, as well as good exercise. Practice perches may be made of ¾-inch (2 cm)-diameter wooden dowels or 1×2 lumber with the narrow edge turned upward. Allow 4 inches (10 cm) of roosting space per bird, making the total length of the perch no more than about 3 feet (90 cm) so it won't sag under the weight of multiple birds.

Either provide low and high perches, or start the perches close to the floor and move them up as the majority of birds learn to use them. If space permits, an additional swinging perch provides baby fowl (as well as their keeper) with endless hours of amusement.

Providing a practice perch for young land fowl helps ensure they will readily roost when moved to their adult quarters.

5

HATCHLING HEALTH ISSUES

Many things can go wrong in brooding poultry, but take heart — the worst things usually happen to birds brooded on a commercial or industrial scale. By brooding in small numbers and maintaining a healthful environment, you're way ahead of the game. However, anyone who broods any kind of poultry long enough will sooner or later experience some of the more common health issues.

The most common health issues among chicken hatchlings, are the three Cs: coccidiosis, cannibalism, and crooked toes. For poults the issues are coccidiosis, cannibalism, and blackhead. Keets are subject to coccidiosis, as well (although mine have never experienced a health issue of any kind). Ducklings and goslings, too, are for the most part extremely healthy and, like all poultry, remain so if fed a balanced diet, furnished a continuing supply of clean water, and housed in a well-maintained brooder of adequate size and appropriate temperature.

Contagious Diseases

Disease is rarely a problem in hatchlings unless they pick up an egg-transmitted illness during incubation, the incubator or brooder isn't kept clean, or the birds are otherwise exposed to a high concentration of typical poultry microbes before they have a chance to develop natural immunity through gradual exposure.

Hatchlings need time to build up their immune systems, and until they do they are fairly vulnerable to any disease they might become exposed to.

They are especially susceptible to adult illnesses, so do not brood babies in unclean housing where mature birds have been raised in the past. When you do chores, take care of your babies before visiting your older birds.

Different poultry diseases occur more or less commonly in different parts of the country. Your county Extension agent or state poultry specialist can tell you whether or not your birds should be vaccinated and for which diseases.

INDICATIONS OF HEALTH

When you are familiar with how a healthy bird looks and acts, you can detect illness easily by noticing changes. Each time you visit your brooder, stand quietly for a few moments and watch for anything unusual.

Smell. Notice how your birds normally smell. Any change in odor is a bad sign.

Sound. Healthy birds make pleasant, melodious sounds. Sick birds may sneeze, wheeze, or make other abnormal sounds.

Appearance. A healthy bird looks sharp and perky when it isn't sleeping. A sick bird may become droopy and inactive, pull in its neck or let its head hang down with eyes closed, have crusty or swollen eyes, appear weak or wobbly, let its wings hang loosely, or simply stop growing.

Droppings. Baby-bird poop is somewhat firm, although occasional thick, sticky droppings are quite normal. Abnormal droppings may be soft, watery, or bloody. Runny poop dribbling down the back end is a bad sign.

Behavior. Notice the amount of feed your birds normally eat and the amount of water they drink. Unhealthy birds may drink more than usual, may eat or drink less than usual, and may stop growing.

A baby bird that doesn't feel well looks droopy and inactive.

COCCIDIOSIS

The most common disease of brooded poultry is coccidiosis, which affects primarily chicks and poults; rarely, keets, ducklings, or goslings. Coccidiosis, or cocci (*coxy*), is an intestinal disease that is most likely to appear in birds that are three to six weeks of age, with the worst cases occurring at four to five weeks. It is the most common cause of death in growing poultry.

Although many different animals may be infected with coccidiosis, the same species of coccidia protozoa that infect poultry do not affect other livestock. The opposite is also true — poultry cannot get coccidiosis from other animals or, with one rather rare exception, from each other.

Cause

Coccidia are tiny parasitic protozoa that naturally colonize the intestines of birds and other animals. Some of them are harmless, while others cause serious illness if they become too numerous. Coccidia may multiply too rapidly if birds pick up a lot of them by eating droppings in the feed,

Talking Fowl

cannibalism. The nasty habit chicks and poults have of eating each other's feathers or flesh, which may occur when they grow their first feathers

contagious. Description of disease that spreads by direct or indirect contact with an infected individual

medicated starter. A starter ration containing a coccidiostat to prevent coccidiosis

pasting. Also called *paste up, pasty butt,* or *sticky bottom;* a condition in newly hatched chicks whereby soft droppings stick to the vent and harden

stressor. Any negative experience, situation, or event that causes psychological or physical tension or anxiety

water, or litter. Litter picking occurs when feed runs out or birds are so crowded they either get bored (because they have too little room in which to move around) or can't get enough time at the feeder (because too little feeder space has been provided for the number of birds).

Coccidia have a direct life cycle: A female protozoa in a bird's intestines sheds eggs that are expelled in the bird's droppings. The protozoa eggs may be eaten by the same bird or by a different bird, in which they hatch into more protozoa. For each egg that hatches in a bird's intestines, millions are later released in the bird's droppings.

Chicks are susceptible to nine different species of coccidia, each of which invades a different part of the bird's intestine. A chick may be infected by more than one species at a time. Chicks become immune only to the species occurring in their environment. Healthy chicks brought together from different sources may not all be immune to the same species and therefore may transmit disease to one another — with devastating consequences.

All ground-fed birds are exposed to coccidia throughout their lives. Gradual exposure allows them to develop immunity. Coccidia thrive in warm, humid weather, so birds reared early in the year while the weather is cool are in the best position to develop gradual immunity as the weather warms. Birds brooded on clean loose litter develop gradual immunity. Birds kept in a brooder with wooden slats or a hardware cloth floor, then later moved to housing with a solid floor, have had little exposure and therefore have no immunity and can become seriously infected.

Poults are also susceptible to coccidiosis, but the protozoa that affect chickens are not the same as those affecting turkeys, and vice versa: chickens cannot transmit coccidiosis to turkeys, and turkeys cannot transmit it to chickens. One species of coccidia infects two different poultry species — guineas and turkeys — but it is quite rare. All other coccidia infect only one species of poultry. Furthermore, keets, ducklings, and goslings are much less susceptible to coccidiosis than are chicks and poults.

Medicated Feed and Waterfowl: Caution

Although ducklings and goslings can get coccidiosis, they rarely do. Never feed baby waterfowl medicated rations designed for chickens or poults. The dosage in medicated feed is calculated based on the expected rate at which the ration will be eaten, and if ducklings or goslings eat more than the calculations allow for, they can overdose on the medication. Medicated starter fed to ducklings or goslings must be formulated specifically for waterfowl. If you can't get it in your area, feed your waterfowl unmedicated chick starter or, if they are being raised for meat, higher protein poult or game-bird starter.

Prevention

When you purchase chicks from a hatchery, you may be offered the option of having them vaccinated against coccidiosis. This stimulates a natural immunity that produces lifetime protection against cocci, but take care to never feed medicated rations, which would neutralize the vaccine.

Coccidiosis may be prevented in unvaccinated birds by the use of medicated starter ration, but the better option is to manage litter to prevent damp spots and caked droppings, and keep feeders and drinkers free of droppings. Wash drinkers each time you refill them. If you find droppings in the water, elevate the drinker on a low platform, or switch to one that's more suitable.

The best measure of all is to get your poultry out of the brooder and onto clean pasture as soon as the weather permits. A properly maintained pastured flock is less likely to become infected than birds living in crowded conditions, brooded on damp litter, or allowed to drink water fouled with droppings. Move pastured birds often enough that they don't have to forage in accumulated droppings.

Newbie poultry owners — as well as anyone who broods chicks or poults late in the season or in a year-round warm climate where coccidiosis is difficult to control — would be wise to use a coccidiostat, especially when brooding in facilities that have housed poultry in the past. This medication, which prevents coccidiosis, comes in either a medicated starter ration or a liquid form that's added to the drinking water. Medicate water with caution; birds drink more in warm weather and can obtain a toxic dose.

A coccidiostat is a preventive measure that won't help birds already infected with coccidiosis.

Treating the disease requires stronger medication, which is available from a farm store, poultry supplier, or veterinarian. For my take on non-drug methods for treating coccidiosis — beyond painstaking daily attention to brooder and coop sanitation — see Screwpot Notions in the appendix.

Control

Chickens and poults that come down with coccidiosis can't be cured with medicated feed, and besides, sick birds typically stop eating. This disease is serious — now is not the time to experiment with any of the numerous cures described on the Internet. Birds with coccidiosis must be treated with an anticoccidial dissolved in their water, since they will continue to drink even after they stop eating. Treatment works quickly and effectively if started at the first signs of disease.

The main signs of coccidiosis are sitting hunched up and rumpled looking; slow growth; and loose, watery, or off-color droppings. If blood appears in the droppings, the illness is serious — birds may survive but are unlikely to thrive. The disease may develop gradually, or bloody diarrhea and death may come on fast. Birds that have been infected and then treated will be immune to further infection from the species of coccidia that caused the infection, but they often do not grow well.

Not all anticoccidial medications work against all species of coccidia, and using the wrong drugs can do more harm than good. If you suspect coccidiosis, isolate sick birds, take a sample of fresh droppings to your veterinarian, and ask for a fecal test to find out what kind of coccidia are involved and which medication to use.

MAREK'S DISEASE

Marek's disease is an extremely common incurable viral infection of chickens, and rarely turkeys, which results in tumors and sometimes death in growing birds. It is caused by six different strains of herpes virus that keep evolving and becoming more virulent.

The virus can lie dormant in infected chicks and may not become apparent until toward the end of the brooding period, or even months later. Meanwhile, it compromises the immune system, making a chick more susceptible to other diseases. Chickens infected with this disease shed the virus whether or not they show signs, thereby contaminating their housing for future chickens.

The most common sign in chicks over three weeks of age is growing thin while eating well, with deaths starting at about eight to ten weeks of age. An older infected chick may recover after temporarily losing control of its muscles and taking on the characteristic appearance of lying on its breast with one leg sticking out straight forward and the other trailing behind.

Prevention includes brooding chicks away from all mature chickens, practicing good sanitation, and vaccinating hatchlings. Vaccination does not prevent chicks from becoming infected and shedding the virus but does prevent paralysis.

Typical posture of a chicken paralyzed by Marek's disease.

For the vaccine to be effective, chicks must be vaccinated before being exposed to the virus. Most hatcheries offer to vaccinate chicks before shipping them. However, because the virus keeps mutating, not all vaccines are effective against all strains of the virus.

Marek's disease is not spread by means of hatching eggs. In fact, some degree of immunity is transferred from hen to chick to protect a chick during its first few days of life. Chicks that are not exposed to this virus or adult carriers until they are five months old usually have developed a natural immunity. Some breeds, particularly Fayoumis, are naturally resistant, while others, notably Sebrights, are more susceptible than most breeds. Brooding chicks with poults confers some degree of immunity to Marek's disease but carries the risk that the chicks may infect the poults with blackhead.

BLACKHEAD

Histomoniasis is a serious and incurable disease of young turkeys. It commonly appears toward the end of the brooding period, or for several weeks thereafter. It is caused by protozoa known as histomonads that live in most poultry environments, except where soil is dry, loose, and sandy. The disease is known as blackhead, which is misleading because the head of an affected bird does not always darken, although sometimes the disease causes a reduction of oxygen in the blood, resulting in the skin's appearing dark blue or black.

Early signs include droopiness, weight loss, and loose yellow droppings. Sometimes the droppings are tinged with blood, which may be mistaken as a sign of coccidiosis, but the color of blackhead droppings is a distinctive sign. Poults die from either liver failure or a bacterial infection following damage caused by the protozoa. Dead poults are sometimes the first sign of this disease.

The protozoa are carried by the cecal worm (*Heterakis gallinarum*), the most common parasitic worm in North American poultry. The infection is spread by cecal worms and their eggs, expelled in droppings, and then picked up by a foraging turkey. Earthworms, too, may eat infected cecal worm eggs, then infect a turkey that eats the earthworms.

Protozoa that are expelled without being protected inside a cecal worm egg quickly die, but given such protection, they can survive in the environment for years. Once turkeys become infected, blackhead is difficult to eliminate from the flock or the land. Since the protozoa require moisture to survive, raising poults in a dry environment helps prevent this disease.

Other preventive measures include keeping poults away from grown turkeys and chickens and away from land where any poultry have lived for at least 3 years. The time period may be shortened to 2 years if the land is tilled and planted to grass or a garden before being returned to a poultry yard. Providing practice perches early encourages poults to roost off the floor, away from potentially contaminated droppings.

Brooding Poults with Chicks

Conventional wisdom says you should never keep turkeys and chickens together, because turkeys are susceptible to the devastations of blackhead. However, lots of backyarders raise chickens and turkeys together without a problem, and with some benefits.

Newly hatched poults tend to get off to a slow start, but chicks are somewhat quicker on the uptake. Poults brooded with chicks learn to eat and drink more readily by following the chicks. And chicks raised with poults acquire a sort of immunity to Marek's disease. Turkeys carry a related, although harmless, virus that keeps the Marek's virus from causing tumors in chickens.

On the downside is the blackhead issue. Although chickens commonly carry the blackhead protozoa without being infected, poults are highly susceptible and can easily become infected by chickens. Since an outbreak of blackhead requires the presence of both histomonads and cecal worms, the danger of blackhead may be minimized by avoiding bringing in additional poultry, which may bring trouble with them, and, where cecal worms are already present, by regularly worming mature poultry to reduce the cecal worm population.

Sudden Death

Many backyarders regularly brood poultry without losing a single bird. Among commercially raised chicks, however, the normal death rate is about 5 percent during the 6-week brooding period. The average death rate for commercially raised poults is about 15 percent, and some growers do much worse. No matter how careful you are, the occasional bird may die. As distressing as a death may be, the random loss of a hatchling is not a cause for panic.

Most hatchlings that aren't going to make it will die during the first week after the hatch. Whenever you visit your brooder, check to make sure none of the birds has died and needs to be removed from the brooder. If you use a heater panel that's difficult to see underneath, lift it up so you can see what's going on under it. A dead hatchling rapidly gets trampled into the bedding and can be hard to find.

Naturally, you'd be upset if you found a dead bird, but it's not a major health issue unless several birds die at once or in succession, or those remaining alive exhibit other signs of disease. Should that occur, isolate the sick ones and seek help from someone experienced with poultry, an avian veterinarian, or your county Extension agent. Online forums frequented by experienced poultry keepers can also be helpful.

Physical Defects

More common than disease issues for backyarders are functional defects, many of which involve the feet and legs and therefore hinder mobility. Happily, most of these problems are preventable. And if they do occur, most of them may be mended easily.

Deformities that typically appear at the time of hatch include crooked toes, splayed leg, deformed beak or bill, and twisted neck. These conditions are discussed in detail in chapter 10.

LAMENESS

The most common functional affliction of waterfowl is lameness, because they have structurally weak legs. Their legs are made for flying and swimming, not walking. Lameness may occur from an injury that causes a sprain (never grab a duckling or gosling by a leg) or from a sliver of some sort lodged in a footpad, in which case you must remove the sliver and disinfect the foot. Such incidences generally involve only one duckling or gosling.

A nutritional deficiency — most commonly, insufficient niacin — affects an entire brood of waterfowl. The birds may decline gradually or may go from getting around fine one moment to suddenly being unable to walk. Feeding a poor-quality waterfowl ration is the culprit. The solution is to immediately switch to a quality brand formulated specifically for waterfowl. If you are feeding starter formulated for land fowl, add 8 ounces (.25 L) of brewer's yeast to each 10 pounds (4.5 kg) of starter. Within a day or two, your ducklings and goslings should be back to scurrying around as usual.

In land fowl brooding on wooden slates or on suspended wire mesh can lead to lameness. The lighter breeds may get a foot or hock caught in a space between slats or wire mesh, and the heavy breeds weigh too much to walk handily on slats or wire.

TWISTED WING

Twisted wing, also called slipped wing or angel wing, is a condition in which one or both wings have one or more twisted feathers. Either the primary feathers overlap in reverse order — over rather than under each other from outer to inner — or, more commonly, the entire last section of the wing flops to the outside, angling away from the body like an airplane wing. This condition may be genetic or may be caused by a dietary imbalance.

It is more often seen in waterfowl, particularly geese, than in land fowl and is more common in ganders than in hens. In geese the deformity occurs usually when flight feathers grow faster than the underlying wing structure. The heavy feathers pull at the wing, causing the wing to twist outward. When the bird matures, the affected wing remains awkwardly bent outward instead of gracefully folding against its body.

Do not confuse twisted wing with a droopy or lazy wing. As birds grow sometimes their little wings have trouble holding up all the newly sprouting feathers, and one or both wings may slightly droop without twisting outward. That's the time to watch for wings twisting outward instead of merely drooping.

Twisted wing is not an issue for waterfowl to be harvested for meat. Neither is it a defect in Sebastopol geese, with their curly feathers; rather, it's a characteristic of the breed. But it is unsightly in other geese and ducks (or any bird) kept for showing, breeding, or just for fun.

Prevent this condition by avoiding excess protein. Letting young waterfowl graze helps. To treat, switch from high-protein starter to alfalfa pellets so the wing's structure can catch up with feather development. Vet wrap securing the last two joints of the wing for 4 or 5 days will hold the feathers in proper position to help the wing grow in the right direction. Remove the vet wrap each night so the bird can exercise its wing muscles.

Drooping Wings Can Twist

When a growing bird has trouble holding up its heavy wing feathers, one or both wings may droop lazily until the bird gets a little older and stronger.

Heavy feathers may pull at the wing, causing the wing to twist awkwardly out to the side. This condition may become permanent unless corrective measures are taken.

Twisted Wing Repair

Vet wrap around the last two joints of a twisted wing will hold the feathers in proper position and help the wing grow properly.

STEP 1

Wrap a 12-inch length of narrow vet wrap around the tip of the wing in the direction shown by the arrows. Then pass the vet wrap under and behind the upper section of the wing, pulling the two sections of the wing close together.

STEP 2

Wrap the remainder of the vet wrap around both sections of the wing to hold them together, taking care not to pull it too tight. The stronger upper section of the wing now provides support for the tip, preventing it from twisting outward. Remove the wrap periodically to give the bird a chance to exercise its wing.

WRY TAIL

I once met a chicken keeper who pronounced *wry* as "weary," which pretty much describes the condition in which the tail leans to one side as if too weary to remain erect. Indeed, wry or twisted tail is caused by weakness in the vertebrae that hold the tail, allowing the tail feathers to flop to one side. It usually appears as the tail feathers grow and become heavy. It is more common in cockerels than in pullets. A bird with wry tail cannot be shown and, because the condition is caused by a genetic defect, should not be kept as a breeder.

Wry tail is caused by weakness in the vertebrae that hold the tail, allowing the tail feathers to flop permanently to one side.

Environmental Sickness

A number of common conditions in young poultry are related to their environment, which is largely an issue of brooder management. For instance, brooded birds that are too hot or too cold will stop eating, which is definitely not good for their health. Other conditions might be related to the placement or sanitizing of feeders and drinkers, or to situations that result in undue stress.

STRESS

Brooded birds are always under stress in one form or another. Excessive stress can reduce their growth rate and make them more susceptible to diseases they might otherwise resist. To avoid introducing more stress, do not handle hatchlings for their first few days of life. Do talk or hum whenever you approach so they won't be startled by your sudden appearance, and move gently when tending the brooder.

The most routine activities can increase stress. To minimize the effects of stress, avoid making more than one change at a time. For example, avoid moving and vaccinating birds at the same time. When feeders or drinkers are replaced with larger ones, leave the old ones in place for a few days until the birds get used to the new ones.

Approaching the birds from the side, rather than from the top, may also reduce stress. Commercial box and battery brooders are designed with this feature in mind. Most homemade brooders are designed for the convenience of the poultry keeper, who can scare the living

Stressed Birds? Add Apple Cider Vinegar

Intestinal microflora prefer a pH range of 5.5 to 7.0; disease-causing microbes prefer a pH range of 7.5 to 9. Stress puts poultry into "fight or flight" mode, diverting metabolic processes from the digestive system to the muscles. As a result their bodies shut down the production of digestive acids, causing intestinal pH to rise and paving the way for an explosion of bad microbes. When brooded birds are subjected to excessive stressors, adding apple cider vinegar to the drinking water at the rate of 1 tablespoon per gallon (15 mL per 3.75 L) — double the dose if the water is alkaline — reduces pH in the crop to encourage microflora to flourish there, ensuring they make it to the gut to keep the little birds' intestinal flora strong and healthy.

Common Stressors

SEVERITY	STRESSOR
Usually minor	Too much time between hatching and first food or water
	Nutritional deficiency in chicks due to inadequate breeder-flock diet
	Cold, damp brooder floor
	Debeaking
	Rough handling
	Low-grade infection
	Eating spoiled feed
	Unusual noises or other disturbances
Moderate	Chilling or overheating during first weeks of life
	Extremely rapid growth
	Chilling or overheating during a move
	Sudden exposure to cold
	Extreme variations in weather or temperature
	Unclean feeders, drinkers, or litter
	Internal or external parasites
	Insufficient ventilation or draftiness
	Vaccination
	Competition between sexes or individuals
	Medication (severity depends on drug used)
Serious	Overcrowding
	Nutritional imbalance
	Insufficient drinking water
	Combining birds hatched more than a week apart
Severe	Suffocation caused by piling
	Lengthy periods without feed or water
	Inadequate number of feeders
	Poorly placed feeders causing starvation
	Onset of any disease

Adapted from: *Storey's Guide to Raising Chickens*, 3rd Edition

daylights out of baby poultry by approaching them from above — after all, most predators swoop down on small birds. If you approach your brooder from the top, don't forget to hum or speak softly to let your birds know you're coming.

Brooding multiple ages or multiple species can be relatively stress free if done judiciously. Birds of the same species that hatch up to a week apart may generally be combined in the same brooder. If they hatch more than a week apart, the older ones will likely be too active for the hatchlings. Poults, chicks, and keets may be brooded together if they hatch at the same time, or within day or two of each other. Even though poults are larger than chicks, and chicks are larger than keets, their overall activity level at the same age makes them compatible. The same is true of ducklings and goslings. Brooding land fowl together with waterfowl, however, is a formula for disaster — waterfowl like to create a wet mess, which is a major stressor for land fowl.

SMOTHERING

Smothering can occur if birds don't get enough heat, or they feel a draft, causing them to pile on top of each other in an attempt to get warm. As a result the ones on the bottom may be trampled or suffocated. Smothering due to drafts or insufficient heat usually occurs at night when the ambient temperature goes down. A cardboard ring encircling the birds (or around the inside walls of the brooder) during their first week of life keeps them close to the source of heat and out of corners.

PASTING

Pasting — also known as pasty butt, paste up or sticky bottoms — is a common condition in newly hatched chicks. Soft droppings that stick to a chick's vent harden and seal the vent shut, eventually causing death.

Although pasting may be caused by disease — typically in chicks older than one week — it is more likely to be caused by chilling, overheating,

Ventilation versus Draft

The ideal brooder provides good ventilation without being drafty. Isn't that a contradiction? Not at all.

Ventilation involves a steady air exchange whereby stale air goes out and fresh air comes in. Brooded poultry need increasingly more ventilation as they grow, because they generate increasingly more moisture from breathing and pooping. Since warm air rises, an area brooder or open-top brooder generally provides adequate ventilation.

Good ventilation doesn't necessarily stir up a breeze, but a drafty condition does. To detect a draft hold out a strip of survey tape or tissue paper down at chick level. If the tape or paper moves while you are holding it still, you have a draft. The way the tape blows shows you where the draft is coming from, so you can find it and fix it. Also, if chicks all face the same direction, there's a draft, and they are facing it.

Smothering also occurs when birds are transported in stacked boxes with too few ventilation holes, or in the trunk of a car where air circulation is poor. Fright may cause piling in growing birds. Ducks are particularly sensitive to moving light, such as from passing vehicles, and all birds become nervous after they've been moved to unfamiliar housing. When you move birds to new housing, keep the lights dim and check often during the first few nights to make sure they are resting normally.

or improper feeding. Shipped chicks that were chilled in transit may paste, as may dehydrated chicks that get too-cold water as their first drink. Pasting is less apt to occur if the first drink is brooder temperature and the chicks are drinking well before they start eating.

Another preventive measure is to feed only chick scratch, or starter combined with chick scratch, for the first day or two, while the chicks are still deriving nutrients from residual yolk. If your farm store doesn't carry chick scratch, run regular chicken scratch through the blender, or crush uncooked oatmeal and combine it with an equal amount of cornmeal.

Pasting can occur if too much sugar is added to the first water as an energy booster. Some types of feedstuffs, particularly soybeans, can also trigger pasting. The simplest solution is to change to a better-quality starter ration. Our chicks persisted in pasting year after year until we switched feed brands. We haven't had a single paste-up since.

Treatment

A chick with a sticky bottom must be cleaned up before the droppings harden and plug up the works. Run a gentle stream of warm tap water over the chick's bottom, then take your time to carefully pick off the mess with your fingers, taking care not to rip out any down and tear the chick's tender skin. Depending on how thick and

Pasty droppings sticking to a chick's vent eventually become dry and hard, dangerously sealing the vent shut unless cleaned off.

hardened the pasting is, you may have to pick off a little at a time, then apply more warm water. When all the droppings have been cleared away, dry the chick's bottom by gently dabbing it with a paper towel. Apply a little Neosporin or Vaseline to protect the affected area and prevent fresh poop from sticking.

MANURE BALLS

In a crowded brooder, litter gets caked with droppings. Globs of litter and droppings stick to a chick's feet, causing balls to form at the ends of the toes. These manure balls can result in crippling or infection, and they often invite toe picking (see page 106). Manure balls may also result from crooked toes (see page 194), which cause a chick to walk unnaturally on the sides of its feet.

Treatment

To remove manure balls, set the chick's feet in warm, not hot, water until the hardened manure has softened, then gently pick off the balls. If the toes bleed after they've been cleaned, dry them and coat them with Neosporin antibiotic ointment. Be sure to correct the condition that caused the problem in the first place: droppings accumulating in the brooder because of either crowding or improper bedding management.

Balls of hardened droppings that stick to the ends of a chick's toes may cause infection unless removed.

CROWDING AND AMMONIA

In a crowded brooder, litter may have a strong odor, usually of ammonia. Severe ammonia fumes cause chicks to sit around with their eyes closed, looking perfectly miserable. The smell of ammonia is a sign that litter needs to be changed more often. The crowding situation must also be corrected to prevent the birds' eyes from becoming inflamed, possibly leading to blindness.

BROODER PNEUMONIA

Caused by a fungus that thrives at warm temperatures, brooder pneumonia affects all poultry. The fungus most commonly grows in loose bedding but can grow in contaminated feed or water. The classic sign of the illness is birds with trouble breathing and poor appetites. Take corrective measures, or birds may die.

Brooder pneumonia has no cure. It may be prevented by thoroughly cleaning and disinfecting the brooder between broods; putting hatchlings only on fresh, clean litter; and avoiding the use of shavings or other types of litter that have been wet — as may happen when rain falls on bales of shavings displayed outdoors at the farm store.

STARVE-OUT

When hatchlings don't eat within 2 to 3 days of hatch, starve-out occurs. The birds become too weak to actively seek food and die of starvation. Starve-out may be triggered by a too-long transit time for shipped chicks, by feeders and waterers placed where the birds can't find them, or by feeders set so high the babies can't reach them. Other causes are placing feeders directly under the heat, and brooding hatchlings on loose bedding that they fill up on before they learn to identify what is edible and what is not. For a discussion on suitable bedding for newly hatched poultry, see First Bedding on page 68 in chapter 3.

To get hatchlings off to a fast start in distinguishing edibles from nonedibles, sprinkle a little starter ration over paper towels or in a paper plate, shoe-box lid, or other low-sided container they can stand in while pecking. Once they start eating well, they should soon move on to eating from the regular feeders.

STICKY EYES

Crusty, sticky eyes in young waterfowl may be the sign of a low-grade infection due to the duckling's or gosling's not being able to periodically submerge its head in water and thereby flush out its eyes and nostrils. A supervised swim once or twice a week helps little birds keep themselves clean. Isolate an infected bird, clean its eyes with eyewash recommended by your vet, and provide a nutritiously balanced diet to reduce susceptibility to infection.

Toxins Affecting Baby Poultry

TOXIN	METHOD OF CONTACT	SIGN
Carbon monoxide	Birds transported in the poorly ventilated trunk of a car	Death
Disinfectant, especially containing phenol	Overused or applied in poorly ventilated brooder	Huddling with ruffled feathers
Fungicide	Feeding birds coated seeds intended for planting	Resting on hocks or walking stiff legged
Pesticide	Used to control insects in or around brooder	Death

DROWNING AND WATERLOGGING

As unbelievable as it may seem, ducklings and goslings can drown. Drowning most often occurs when they get into a container of water they can't get out of. They may jump or fall into a kiddie wading pool, for instance, and by the time they've finished splashing around, the water level is too low for them to climb back out. Water that's contaminated with chemicals, including swimming pool chlorine, oil, sludge, muck, or mud, can also lead to drowning. Prevention involves ensuring your ducklings and goslings have no opportunities to swim in foul or chemically treated water or in ponds, pools, or basins that do not provide easy exit. Better yet, never let baby or growing waterfowl swim unsupervised.

The most common cause of drowning among land fowl is an overcrowded brooder, so some birds end up falling asleep against (or in) the drinker. Similarly, a too-hot brooder can force birds to press away from the heater and into the water trough.

The littlest of bantams may climb or fall into a drinker and drown. I have brooded hundreds of bantams and never had one drown, but I have heard from other bantam keepers who are not so fortunate. Sebrights, Old English Game bantams, tiny Seramas, and sometimes keets are particularly susceptible to drowning in a drinker. A preventative measure is to place marbles, or thoroughly cleaned gravel or pebbles, inside the rim of the drinker to reduce the water depth but still allow banties to get a drink. After a week or less, the birds should be coordinated enough to stay out of harm's way.

Waterlogging does not require water deep enough to drown in and can affect waterfowl as well as land fowl. The birds splash so much water around the brooder and get themselves so thoroughly soaked that their downy coats can no longer keep them warm. As a result they can chill and maybe even die. I once found all my brooded keets lying in a wet heap and partially encrusted with starter. I have no idea what possessed them to engage in a water-and-food fight, and never before or since have I seen land fowl of any kind make such a mess.

The rescue procedure is the same for all species: Rinse off the babies one by one in warm, not hot, water. Dry them thoroughly with a terry towel. Put them in a warm, dry place, preferably back into the dried-out and relittered brooder. Cross your fingers that they will all make it. Miraculously, my keets did.

PREDATORS

Baby poultry are especially vulnerable to predators, such as rats and snakes that can sneak through the smallest cracks. I once found a big black snake in my brooder, with three lumps indicating he had eaten three of my chicks. He was too fat to get back out the chicken wire, so he curled up under the heater and went to sleep. Thereafter, we exchanged the chicken wire for hardware cloth, which has a mesh small enough to exclude snakes as well as rats.

Raccoons can pry open doors that aren't latched shut. And don't overlook family pets, which should never have unsupervised access to brooding birds. Typical predators and the clues they leave behind include these:

Cat. Chicks missing, no clues; or birds pulled into brooder wire and eaten, except for wings and feet

Opossum. Chicks missing, no clue

Raccoon. Chicks missing, no clue, or some left behind dead; bits of coarse fur clinging to the opening through which the raccoon came; returns every 5 to 7 days

Rat. Leaves bruises and bites on legs; may pull partially eaten chick, head first, into a tunnel opening; may succeed in pulling entire chick down the tunnel and leave no clues

Skunk. Kills chicks and eats abdomen but not muscles and skin; may leave a lingering odor

Snake. Chicks missing, no clues, unless the snake curls up nearby to sleep off the meal

Weasel. Chicks missing, no clues, except possibly a faint skunky odor

Most predators of baby poultry prowl at night or in early morning. House cats and snakes are an exception; they may nab a chick at any time of day or night. A prowling predator frightens baby birds, possibly causing them to stampede and trample one another. Finding trampled birds in a brooder, with no other evident cause, is a good reason to watch for a possible predator.

Baby poultry are so small and easy to eat that other than having birds missing you may find no clues except possibly the size of any access opening into the brooder in relation to the size of the potential predator. A neighbor once came to tell me his ducklings were eating each other. I explained to him that ducklings don't do that and suggested that a weasel or mink was periodically visiting his brooder. He knew I was wrong because his brooding pen was tight; the only explanation for a missing duckling had to be that the others ate it. Exasperated, I told him he'd find out I was right when he lost the last duckling, which obviously could not eat itself. And that's exactly what happened. Eventually, the neighbor spotted a weasel in his brooding pen, looking for yet another duckling. Although cannibalism might be mistaken for the work of a predator, missing birds should not be mistaken for cannibalism.

CANNIBALISM

Chicks, and sometimes poults, may develop the nasty habit of picking at one another. Picking usually starts with one bird and spreads to others. The two most common forms of cannibalism in the brooder are toe picking and feather picking.

Toe picking usually starts among hatchlings that are just beginning to peck. Once they get the hang of it, they eagerly look around for things on the ground to peck at. Among the things they find are toes — their own toes and the toes of other birds. A common trigger for toe picking is running out of feed to peck at or not having enough feeder space for the number of birds. Make sure the chicks don't outgrow their baby feeders or that replacement feeders are not too high for them to reach. Keep the chicks active by moving feeders far enough from the heat source to encourage exploration.

Feather picking usually starts when chicks begin feathering out. The newly forming pinfeathers contain a supply of blood to nourish the growing feather (hence their other name, blood feathers). In using its beak to groom its feathers, a bird may accidentally pull out one of these tender

Feather picking usually appears when chicks begin feathering out, and commonly starts at the base of the tail.

Picking versus Pecking

Cannibalistic picking is entirely different from peck-order fighting to establish dominance, although frequent fighting to adjust the peck order may lead to bloody injuries, which in turn lead to picking and cannibalism. As birds grow and establish a peck order, regularly introducing new birds disrupts the peck order and may provoke cannibalism.

What Triggers Picking?

Cannibalism has many triggers, often working in combination. Conditions that can trigger this undesirable activity include the following:

- Crowding
- Excessive heat
- Bright light
- Inadequate ventilation
- Empty feeder
- Too few feeders and drinkers
- Feed and water stations too close together
- Diet too low in protein
- Brooding on wooden slats or wire mesh

feathers. Furthermore, these interesting-looking dark feathers sprouting among baby down attract the curiosity of other chicks, which check them out the only way they know how, by pecking. When a pinfeather is pulled out, it spurts a small amount of tasty blood, which attracts more attention and more picking, especially around the tail and along the back. A snack of blood is particularly attractive at a time when feathering birds crave additional protein, enticing chicks to pick their own pinfeathers as well as each other's.

Prevention

Cannibalism may be prevented almost entirely with careful brooder management. The most important thing to do is ensure your birds always have plenty of room to grow, which likely means periodically increasing their living space. Fast-growing birds are especially quick to fill the available space, and when they can't get away from each other, they get frustrated and start picking.

After the first 48 hours, when baby birds are getting acclimated to the brooder environment, do not leave on bright lights 24 hours a day. Use dim lights or at least provide for some dim areas, especially under or near the heater where the birds sleep. Bright lights left on 24/7 don't allow birds time to rest, making them just as irritable as anyone who doesn't get enough sleep.

Regularly reduce the brooding temperature by 5°F (3°C) per week, or more if the ambient temperature is higher than the necessary brooding temperature. Increase ventilation as chicks get older and need more oxygen. Being short of fresh air can shorten tempers.

Remove from the brooder any bird that is ill, lame, or not growing well, as chicks tend to peck at weaker birds. Once they start growing, do not introduce birds of different ages into the brooder, since the stronger, older birds may peck the weaker, younger ones.

Protein requirements increase as birds grow, and protein deficiency can lead to cannibalism — after all, feathers and flesh are good sources of protein. Take care to adjust rations as needed to ensure your birds get adequate protein. Good-quality commercial rations usually contain sufficient protein.

Make sure you have enough feeders and drinkers for the number of birds you are brooding, so they all have plenty of opportunities to obtain feed and water. Switch to larger feeders and drinkers as the birds grow, so they never run out during the day and start picking each other instead of pecking for food. Periodically move feed and water stations farther from the heater, and farther apart from each other, to keep the birds actively seeking feed and water.

This 4-inch diameter Treat Ball, from Happy Hens Treats, when filled with cabbage leaves or other greens, helps relieve brooder boredom.

In general provide a rich environment that encourages foraging, exploring, and other normal behaviors. Brooding on wooden slats or wire mesh can trigger picking because the birds don't have much else to do. Brooding on loose bedding, or at least including an area suitable for dustbathing, offers an opportunity to engage in this pleasurable activity to while the time away. Furnishing practice perches gives birds another way to entertain themselves and additional space to get away from each other.

Offer your chicks and poults something to peck on: a shiny aluminum pie tin hanging from a chain or string; a chunk of wood or a mini bale of hay for them to hop up on and peck at; a flake of hay or a fresh head of cabbage or lettuce suspended where they must reach up to peck it. (Be sure to remove uneaten produce before it rots.) For the entertainment of confined birds, pet stores offer all sorts of toys, some of which are just as much fun for brooded poultry as for cage birds to play with.

Control

Despite your best efforts, brooded chicks or poults will sometimes start picking at each other. Some breeds and strains are simply more likely than others to do so. Light, high-strung breeds, for example, are more likely to pick than the heavier, more sedate breeds. If you have a problem with a particular breed or strain, avoid perpetuating the problem by not hatching eggs from birds that pick each other's feathers or otherwise behave aggressively toward each other.

Remove offenders. If picking does get started in your brooder, the first thing to do is identify and remove any instigators. Also remove birds that have bloody wounds or patches that have been picked bare.

Seeing red. Switch to red lights that make blood more difficult to see, or change from bright lights to dim lights. Provide only enough light to let you barely read a newspaper.

Room to roam. Increase the available living space, preferably by moving the birds to a bigger brooding area or, best of all, by putting them on pasture if they are old enough and the weather is

nice enough. Letting them run outside alleviates boredom by giving them new ground to explore, where they can find things to peck besides each other.

Salt solution. Salt deficiency can cause a craving for blood and feathers. Try adding one tablespoon of salt per gallon of water in the drinker for one morning, then repeat the salt treatment 3 days later. At all other times provide plenty of fresh, unsalted water.

Prepared treatments. Pine tar, menthol ointment, diaper rash cream, and various antipicking preparations have been recommended or devised for smearing on wounds to make them less palatable for picking while they heal. Some of them work for a short time, especially when picking first starts, but none works well after picking has become serious.

Protective specs. Blinders, specs, or so-called peepers are sometimes used to prevent or control cannibalism by keeping birds from seeing directly ahead to aim a peck. Such devices can lead to eye disorders, however.

Beak trim. In commercial operations chickens have the tips of their beaks removed so they may be crowded together without eating each other. Although those chickens are permanently debeaked, brooded chicks that persist in picking may be temporarily debeaked by using nail clippers to remove one-fifth of the upper half of the beak, taking care not to crush the beak. A beak that is properly trimmed should grow back in about 6 weeks.

LINE WHERE CUT IS MADE TO REMOVE BEAK TIP.

Chicks that persist in picking may be temporarily debeaked by removing one-fifth of the upper half of the beak, which should grow back in about 6 weeks.

Part 2
The Eggs

6

THE BROODY HEN

Once you have mature chickens and want to hatch more eggs, the easiest way is to let one of your hens do it. Hatching eggs under a hen offers the advantage of being less time-consuming than using an incubator and is often more successful. The hen handles all the logistics of temperature and humidity control, turning the eggs at exactly the right intervals and keeping the hatchlings safe and warm. You don't have to worry about a thermostat malfunctioning, a dried-out water pan, the power going out, or some klutz tripping over the electric cord and pulling the plug out of the wall.

Nature's Way

For these reasons a lot of chicken keepers prefer to let mother hens handle all their incubating and brooding. Some, in fact, keep hens of a breed known to be particularly broody, specifically and solely for the purpose of hatching eggs laid by other breeds or species.

To hatch and brood poultry nature's way, you need two things: fertile eggs and a broody hen.

If your flock includes a male, you might hatch some of the eggs laid by your own hens. Chapter 7 includes an in-depth discussion on obtaining and handling eggs suitable for hatching. Depending on which breeds you decide to raise, having a hen go broody may or may not be problematic.

Broodiness

Some hens make great mothers. Others are iffy. Still others are completely unreliable. And some are downright dreadful. Over the years I've seen many hatches successfully brought off a nest by chickens, turkeys, guineas, ducks, and geese. I've also seen geese squabble over hatchlings until they trampled them all, turkeys that stayed on the nest so long their hatchlings starved, and hens that abandoned the nest just as their eggs were about to hatch.

You can never tell for sure if a particular hen will make a good broody until she does it. But you can get a pretty good idea based on the past records for the breed as a whole. Aseels, Modern Game Bantams, and Silkies, for example, are particularly known for easily going broody and having the tenacity to see it through to a successful completion. And a hen that has once proven her reliability is likely to repeat it in the future.

Because laying stops when setting starts, throughout the ages people who kept poultry

Talking Fowl

artificial insemination. The introduction of semen into a hen other than by natural mating

brood. A batch of chicks that hatch together

brood patch. A large defeathered bare area on a setting hen's breast

clutch. A batch of eggs or brood of chicks that hatch together

cull. To remove and dispose of an undesirable individual (for instance, an egg, hatchling, or breeder) from the population at large. Also, the individual thus removed

fertile egg. An egg that has been inseminated, making it capable of producing a chick when incubated under appropriate conditions

set or brood. To gather eggs in a nest and sit on them until they hatch

A found object of appropriate size — even an old computer shell — may be upcycled to create a brooding nest, keeping even the fussiest of hens happy and secure.

primarily for their eggs have culled persistent broodies. As a result, the breeds best known for superior laying ability, especially those developed for commercial production, are less apt to brood than others. So don't expect hens of the light, flighty laying breeds to brood, and if they do, they generally are not reliable.

At the other end of the brooding scale are backyard breeds like my cute little Cochin bantams that laid a scant few eggs each spring and promptly went broody. But a hen that gets broody won't necessarily continue for the full 21 days. Some hens lose interest midway and simply walk off. Sometimes a pullet just hasn't gotten the hang of it yet, but even mature hens, like some humans, may not have the attention span to stick with it until the job is done and therefore cannot be trusted with valuable hatching eggs.

Hens of the heavier chicken breeds tend to make good broodies that can handle large numbers of eggs, but a really heavy hen with a loaded nest may break some eggs. Breeds with heavy leg feathering are likely to brood but are not always successful because their stiff leg feathers may flick eggs out of the nest and bowl over newly hatched chicks. This issue may be resolved by clipping the broody's feathers close to her leg.

Some breeds rarely go broody within their first year but may brood successfully during subsequent laying cycles. This deferred broodiness trait is typical of Chanteclers and Fayoumis. Individual hens of other breeds may defer broodiness, as may hens resulting from a cross between a hen from a breed that typically broods and a cock from a non-brooding breed.

Even among breeds that do not typically brood, some strains have stronger brooding instincts than others, and among those with the least brooding instinct, the occasional hen takes a notion to hatch out some chicks. And just as broodiness has been bred out of certain strains, this trait may be improved in a flock by keeping as future breeders birds resulting from a natural hatch.

Guinea hens typically are tenacious broodies but can be careless mothers unless they are confined until the keets get big enough to keep up with the adults. Among turkeys the broad-breasted breeds rarely brood, and when they do their eggs are likely not fertile, as the toms are too big and clumsy to breed naturally; industrially, they are bred by artificial insemination. However, most other turkey breeds make willing setters. The following table indicates the most typical brooding scenario for different chicken, duck, and goose breeds.

Brooding Instinct

BREED	BROODINESS
Ducks	
Ancona	Good–Fair
Appleyard	Variable
Aylesbury	Variable
Buff	Good–Fair
Campbell	Fair–Poor
Cayuga	Good–Fair
Magpie	Good–Fair
Muscovy	Variable
Pekin	Variable
Rouen	Variable
Runner	Fair–Poor
Saxony	Variable
Swedish	Good–Fair
Welsh Harlequin	Fair–Poor
Geese	
African	Good–Fair
American Buff	Good
Chinese	Poor
Embden	Good–Fair
Pilgrim	Excellent
Pomeranian	Good
Roman	Good
Sebastopol	Poor
Shetland	Excellent
Toulouse	Good–Fair
Toulouse, dewlap	Fair–Poor

Brooding Instinct

BREED	BROODINESS	BREED	BROODINESS
Chickens		Chickens	
Ameraucana	Good–Fair	Langshan	Good–Fair
American Game Bantam	Excellent	Leghorn (layer strain)	Poor
Ancona	Poor	Leghorn (traditional)	Good
Andalusian	Poor	Malay	Good–Fair
Araucana	Good–Fair	Marans	Fair
Aseel	Excellent	Minorca	Poor
Australorp	Excellent	Modern Game	Excellent
Barnevelder	Fair–Poor	Naked Neck	Good–Fair
Belgian Bearded d'Anvers	Good–Fair	Nankin	Excellent
Belgian Bearded d'Uccle	Good	New Hampshire	Variable
Booted Bantam	Good	Norwegian Jaerhon	Poor
Brahma	Good	Old English Game	Good
Buckeye	Good	Orloff	Poor
Campine	Poor	Orpington	Good–Fair
Catalana	Poor	Penedesenca	Poor
Chantecler	Deferred	Phoenix	Good
Cochin	Excellent	Plymouth Rock	Good
Cornish	Good	Polish	Poor
Crevecoeur	Poor	Redcap	Poor
Cubalaya	Good	Rhode Island Red	Fair–Poor
Delaware	Good	Rhode Island White	Poor
Dominique	Good	Rosecomb	Poor
Dorking	Good	Sebright	Poor
Dutch Bantam	Good	Serama	Poor
Faverolle	Good–Fair	Shamo	Good–Fair
Fayoumi	Deferred	Sicilian Buttercup	Poor
Hamburg	Poor	Silkie	Excellent
Holland	Fair–Poor	Spanish	Poor
Houdan	Fair–Poor	Spitzhauben	Fair
Hybrid layers	Poor	Sultan	Poor
Japanese Bantam	Excellent	Sumatra	Excellent
Java	Excellent	Sussex	Variable
Jersey Giant	Excellent	Welsumer	Fair–Poor
Kraienkoppe	Variable	Wyandotte	Good–Fair
La Flèche	Poor	Yokohama	Good
Lakenvelder	Poor		

SHE'S READY TO START A FAMILY IF . . .

Most hens of any species, but especially the seasonal layers, go broody in spring or early summer. Some chickens may brood in late summer or early fall, and a few may brood at any time of year. Just because a hen is sitting on a nest doesn't necessarily mean she's setting. She may still be thinking

A broody hen disturbed on the nest will puff out her feathers, peck at your hand, or threaten you with growls.

about the egg she just laid, or she may be hiding from some bully that's higher in the peck order. But if she remains on the nest overnight, that's a pretty good sign.

To test a hen for broodiness, gently reach beneath her, as if to remove any eggs you find there. If she runs off in a hysterical snit, she's probably not broody. If she puffs out her feathers, pecks your hand, or threatens you with growls, things are looking good. Within 2 or 3 days, she'll likely settle down to serious business.

Among chickens, clucking is one sure sign of broodiness. Many hens won't cluck until their eggs are ready to hatch, but some start clucking almost as soon as they start setting. The hen's habit of clucking to reassure her chicks has led to the nickname "clucker" for a broody hen.

Once you have determined the hen is serious, mark your calendar so you'll know when to expect the hatchlings to appear. The incubation timetable on page 175 lists average incubation periods for different species. Just for the record, if a hen incubates eggs other than her own, it's the species that laid the eggs, not the one incubating them, that determines how long they will take to hatch.

Featherless Brood Patch

A seriously broody hen loses feathers from her breast to line her nest. In doing so, she develops defeathered brood patches. These bare spots serve two purposes: they keep the eggs from drying out by lending moisture from her body, and they bring her body warmth closer to the eggs.

Although some heat passes through the egg, the top next to the brood patch is somewhat warmer than the bottom next to the nest. To even things out the hen periodically rearranges the eggs beneath her. She also occasionally leaves the nest, during which the eggs cool down.

Although artificial incubation attempts to imitate this process, it rarely results in the same successful hatching rate or in chicks that are quite as strong and healthy. From this perspective, compared to artificial incubation, a broody hen wins hands — or brood patch — down.

INDUCING BROODINESS

The hen, not you, decides when she's going to set. Broodiness is triggered by the hormone prolactin, released by the pituitary gland usually at the time of year when day length is increasing and the hen has more access to greens than to grains. Aside from injecting her with hormones, no foolproof way has been devised to *make* a hen brood. Occasionally, someone will get the bright idea of trying to encourage a reluctant hen to set by enclosing her in a small coop or cage with a pile of eggs. The usual result is a mess of scattered eggs, smashed up by a frantic hen trying to get out.

The best way to find out if a hen will go broody is just to let eggs accumulate in the nest. Among hens that tend to brood at the slightest whim, the sight of a few eggs in a nest is enough to set them off. Because you don't know exactly how long that might take, instead of leaving good hatching eggs in the nest, start with half a dozen eggs of lesser value, or with plastic or wooden eggs from a hobby shop. Marking old eggs with a felt pen or china marker will let you readily distinguish them from freshly laid eggs, which you might wish to continue removing from the nest. After you have determined that the hen is serious about setting, remove the old or fake eggs and replace them with good hatching eggs.

Meanwhile, collect and store the eggs you wish to hatch (as described in chapter 7) for return to the nest later. By storing the selected eggs under optimal conditions for hatching, you can always revert to artificial incubation should none of your hens be willing to hatch them for you. Of course, if you prefer to take a chance on letting nature take its course, just let eggs accumulate in the nest and see what happens. The worse that can happen is that no hen will go broody and the eggs will become too stale to hatch.

CLUTCH SIZE IS RELATIVE TO HEN SIZE

We humans tend to believe hens lay eggs so we can eat them, and to that end over the centuries we have extended the laying season by increasing clutch size, the number of eggs a hen lays consecutively without taking a break. But the real reason poultry lay eggs is to make more poultry. Poultry are, after all, just birds. And like all birds, hens originally found a suitable site for a nest, laid a clutch of eggs there, and went about hatching them. The size of a clutch was ideal for reproductive purposes — small enough for a hen to cover with her body but big enough to ensure that at least some of the hatchlings made it to maturity.

A hen of any species or breed can cover approximately a dozen eggs of the size she lays. If she's going to hatch eggs that are larger than her own (as when a chicken foster mothers duck eggs) don't expect her to handle more than about 10, and if they are much larger than her own (as when a chicken is hatching turkey or goose eggs) maybe 5. If the eggs are smaller than her own (for example, when a chicken hatches bantam or guinea eggs), she might be able to cover as many as 18.

At any rate, all the eggs must fit handily beneath her. If any stick out around the edge, chances are she'll eventually rotate them back under her and let some other eggs take a turn getting chilled, until they all fail to hatch.

Persistent Broodies

Among chickens the best setters may hatch several broods a year, especially if you take away their hatchlings to raise in a mechanical brooder. Such a dedicated hen needs time off between clutches. A setting hen eats about one-fifth of the amount she normally eats, and on some days she won't eat at all. While she's setting she'll lose as much as 20 percent of her normal weight. At that rate a persistent broody that hatches clutch after clutch without a break could eventually starve to death. For this reason some chicken keepers discourage their hens from brooding more than once a year.

If you allow eggs to accumulate in a nest, mark each egg with the date on which it is laid. That way, if the nest gets too full of eggs by the time the hen starts setting, you will be able to remove the oldest eggs so she'll always have the freshest eggs to incubate.

All the eggs in a clutch must hatch at approximately the same time so the hatchlings can leave the nest together, under the protection of the mother hen. Therefore, embryo development within an egg is suspended until a clutch is complete, brooding commences, and warmth from the hen's body causes the embryos to begin developing.

Average Clutch Size

SPECIES	NUMBER OF EGGS
Chicken	12–14
Bantam	8–10
Duck	12–18
Goose	6–12
Guinea	24–30
Turkey	12–18

Brooding Facilities

Brooding requires a quiet place where a hen can calmly go about her business without being disturbed and where she will not be vulnerable to predation. Providing such a place for a chicken is relatively easy, because most hens may be moved after they start brooding. Moving the nest of a guinea, turkey, duck, or goose is much more problematic, and therefore greater care is needed in establishing safe nesting sites.

BROODY CHICKENS NEED PRIVACY

Separating a brooding hen from the rest of the flock is a good idea for many reasons.

Nest switching. When a broody hen gets off her nest to eat, another hen may decide to enter her nest to lay an egg. The returning broody, on finding her nest occupied, may opt to settle into a different nest containing eggs other than those she already started incubating. And if the two birds didn't start brooding at the same time the incubation period will be shortened for one hen and prolonged for the other, with the potential that the latter may leave her nest before the eggs hatch.

Nesting Specs

Regardless of species, a suitable brooding facility has the following characteristics:

- It is well ventilated
- It provides protection from wind, rain, sun, and temperature extremes
- It offers protection from predators
- It is isolated from other poultry.
- It may be darkened by placing it in a dark corner or facing it away from light.
- It has a nest box large enough for the hen to fit fully inside and turn around in and tall enough for her to sit upright without bumping her head.

This chick brooder is roomy enough to house a broody hen, complete with a feeder and drinker, with plenty of space for hatchlings.

Incubation interruption. If other hens continue laying eggs in the broody hen's nest after setting commences, the broody hen may not be able to cover and warm all the eggs, and as a result some or all of the embryos may die.

Chick attack. Assuming a hen successfully completes the hatch, other chickens may kill the fuzzy intruders, much as they would destroy a mouse or a frog wandering into their digs.

Even if you have two or more hens brooding at the same time, ideally, they should be separated from each other. Otherwise, the hens may pile together into one nest and abandon any others. But even if no nest switching or overloading takes place, all the hens may follow the first chicks that hatch, leaving the remaining eggs to chill. Or, assuming all the hens successfully bring off their broods, one may commandeer the feed and water stations for her chicks and chase the others away.

All these potential problems may be avoided by separating your broody hens, each in her own brooding compartment. In my barn I have a series of 3-foot-by-4-foot heated brooders designed for raising chicks, but with the heaters turned off, they are equally suitable for housing broody hens. Each unit can accommodate a hen nesting in a back corner and a feed and drinker toward the front, with plenty of room for a number of hatchlings during their first few weeks of life.

Brooding nests need not be elaborate. A covered kitty litter box or a pet carrier with the door removed makes a perfectly acceptable brooding nest and is easy to move to a private area once a

Moving a Broody Hen

Chickens are easiest to move, but any hen may be moved successfully if:

- She's been brooding seriously for several days
- She's moved after dark
- She and her eggs are moved with all possible swiftness
- Once moved, she's left undisturbed until she's fully settled

An experienced broody hen is often easier to move than a hen that is brooding for the first time. Move the hen at night so she's less apt to try to get back to her old nest. A hen that's brooding for the first time is more likely to stick with it if you move the entire nest along with her. For this purpose a portable nest such as a pet carrier is ideal. If she has not built her nest in a pet carrier, you might be able to induce her to do so by moving her nest into the pet carrier and putting the carrier in the spot she originally chose. If she accepts the nest in the carrier, you can then move the whole thing to private quarters. To avoid disturbing her the next morning, have feeders and drinkers in place and filled with enough food and water to last at least a full day.

hen takes up residence inside. An adequate size for a nest is 14 inches (36 cm) square and at least 16 inches (41 cm) high, with a 4- to 6-inch (10 to 15 cm)-high sill at the front to hold in nesting material and keep eggs from rolling out.

BROODY TURKEYS

Turkeys can be decidedly fickle about where they brood. A hen might plop a bunch of eggs in the corner of a coop and call it good. Or she might fly over the fence and hide her nest in shrubbery or tall grass, where she will be vulnerable to predators. By providing several nesting options within the security of your turkey yard, you increase the chance that your turkey hens will brood in a site of your choosing.

Although this turkey hen appears to be well hidden, nesting in an open field makes her vulnerable to predators.

Possible options include a large pet carrier with the doors removed, a doghouse, or an unused camper shell. The latter offers plenty of room for a nest, along with food and water, and may be closed up at night against predators.

We set up a camper shell in our turkey yard and thought it was a pretty ideal nesting site. Some of our hens agreed, but one felt otherwise. Beneath the overhang behind our barn, we had stored some wire cages, one on top of the other, and one of our turkey hens insisted on laying her eggs on the top cage. When we couldn't deter her, we built a wooden frame to keep the eggs from rolling off and added some nesting material within the frame. Although wild turkeys nest on the ground, our domestic hen was perfectly happy

Unlike her wild kin that prefer to nest on the ground, this Bourbon Red hen chose to nest on top of a stack of hay bales.

with her aerial arrangement. Similarly, we've had turkeys nest at the top of hay bales stacked in our barn.

BROODY GUINEAS ARE A BIT TRICKY

Guinea hens can be more problematic than turkey hens. For many years ours flew off into the woods to build their nests, often never coming back, having met up with a coyote or fox or other hungry critter. But over several generations our guineas have gradually become more domesticated, to the point that some of the hens now nest in a corner of the chicken coop.

A guinea typically accumulates 20 or 30 eggs before she starts setting, and two or three hens may brood together. As a result, the nest gets disorganized and the eggs tend to scatter, so it's always a surprise when keets appear.

Guinea hens are active mothers, and when they nest outdoors their hatchlings have a hard time keeping up; some (or all) invariably get lost. When they nest inside our coop, the hatchlings are confined by high doorsills until they are big enough to fly onto the sill to get out, by which time they're also big enough to keep up with the grown-ups.

Despite their carelessness with their hatchlings, guinea hens are fierce mothers. When they hatch a brood inside a coop that is occupied by chickens or other poultry, they bully the other birds in a misguided attempt to protect their keets. Both the guineas and the other birds would be

When guinea hens accumulate 20 to 30 eggs and attempt to brood together, the result can be chaotic.

These two guinea hens chose a salt-block pan for their joint brooding effort, and its high sides keep their eggs from rolling away.

better off if guineas could be moved to private quarters as easily as chickens can, but once a guinea's nest is disturbed, she may abandon it and establish another site, which may not be easy to find. The only successful way to move a guinea's nest is to do it super swiftly and after dark.

Another way to minimize the problems of lost keets and parental aggression toward other poultry is to let the hens brood in their nest of choice, then remove the keets after the hatch. This system, however, can present a problem for keets brooded apart from adult guineas: when they become big enough to join the adults they will be bullied, or even run off, as intruders.

BROODY WATERFOWL

If your waterfowl shelter is large enough, you might partition off a section to provide individual brooding nests or install pet carriers with the doors removed. However, waterfowl generally do not like to nest in close proximity to one another. Ducks and geese prefer shelters scattered around the yard, so they can investigate each and choose one.

Provide one nest per female. If nests are in short supply, two females may brood side by side, and the resulting hatchlings may get trampled in a squabble over which hen is their mother.

A suitable shelter would be similar to a doghouse without a floor — smaller for ducks, larger for geese. A nest against the ground is better for waterfowl than one with a floor, since soil helps retain the moisture necessary for a successful hatch. Predators, however, can tunnel through the soil to get at the eggs or the female. Hardware cloth secured to the shelter bottom, then topped with a layer of soil, ensures both good humidity and protection from digging predators. Buckets or barrels also make good nests. Block the sides to prevent rolling and tamp clean soil into the bottom for a level floor.

When a brooding duck or goose returns from a swim and hunkers down on her nest, moisture from her feathers keeps her eggs from drying out. As she settles back into the nest, she reorganizes the eggs by rolling them with her bill or paddling

A good way to protect keets from careless mothers, while protecting other poultry from ferocious mother guinea hens, is to gather up the newly hatched keets and put them in a brooder with close-mesh wire on at least one side, and locate the brooder where the mother hen has ready access. Inside the brooder the keets will be safe from being traipsed around the countryside. Meanwhile, the mother (along with her family and friends) will spend a lot of time hanging around the open wire, so when the keets get big enough to be turned loose, the whole family will be just as well acquainted as if the keets had been raised by their mama.

them with her feet. Sometimes during this process an egg rolls out of the nest. A good broody will roll the egg back in with her bill, but a lazy mother will let it go and may thus lose eggs one by one until none is left to hatch. To avoid this problem, add a sill at the front of the nest that is low enough for the hen to step over but high enough to keep eggs from rolling out.

A duck or goose likes to nest where she is protected from the back and sides and can watch for intruders approaching from the front.

Waterfowl like to be protected from the back and sides, so they can face a single entrance to keep an eye out for potential intruders. They also like to brood in darkness and seclusion. Piling brush and orchard prunings in front of the doorway will both darken the nest and camouflage the entry to deter predators. If the weather is sunny and hot, place the shelter under a tree or pile brush on top (or both) to shade the roof to keep things cooler inside.

Once a duck or goose has chosen a nesting site, she will usually continue laying there as long as it remains undisturbed. If one day she enters to find her eggs gone, she is likely to look for a different site. If you wish to hatch more ducklings or goslings than your waterfowl can hatch naturally, collect some of the eggs to hatch in an incubator. Mark a couple of the early eggs, and leave them in the nest. The hen will keep coming back to lay more. After you have collected your share of the continuously laid eggs, you can let a duck or goose accumulate a nest full to hatch herself, or you might help her along by adding another hen's eggs to her nest. Don't forget to remove the old ones you marked, because they will rot in the heat under her breast.

NESTING MATERIAL

Nesting material gives a hen something soft to sit on and to hollow into a bowl for the eggs to nestle in, helping prevent breakage and retain warmth. When a hen leaves the nest to grab a bite to eat, she may cover her eggs with some of the nesting material to keep them warm and make them less visible. If no nesting material is provided, a dedicated hen will gather whatever she can find.

Given the opportunity, the hen will shape her own nest out of clean, dry wood shavings (other than cedar), dry leaves, straw, or hay, always augmented with feathers pulled from her breast. The nesting material must be dry to prevent mold, which can infect developing embryos within the eggs, as well as hatchlings as they emerge from the eggs.

Sometimes eggs break in the nest, either because a hen tries to cover too many or because an egg had a defective shell. Broken eggs attract bacteria and ants, so check the nest occasionally (preferably while the hen is taking her daily stroll), and if necessary, remove any yolk-smeared eggs and replace egg-soaked litter with fresh, clean nesting material. Watch also that the hen doesn't accidentally poop in the nest, which likewise requires cleanup.

Given the opportunity, the hen will shape her own nest out of clean, dry bedding, along with feathers pulled from her breast.

Bug-Free Nests

Be sure both the nesting material and hen are free of lice and mites, which can make a hen restless enough to leave the nest. These body parasites can take enough blood to kill a setting hen or her hatchlings. Old-timers put cedar shavings, tobacco leaves, or mothballs in nests to discourage parasites, but such items are now determined to be toxic to poultry as well as to parasites. Diatomaceous earth or a poultry-approved insecticide sprinkled into the nest before placing the nesting material is a better option.

Managing the Broody Hen

A hen that's brooding in a safe, secure environment requires little by way of management, other than seeing that she has ready access to feed and fresh water and occasionally checking the nest for problems. Place feed and water near the nest, but don't worry if the hen doesn't eat for the first few days. After that she should get off the nest for 15 or 20 minutes (or up to an hour for waterfowl) at about the same time each day to eliminate, grab a few kernels of grain, and maybe take a quick bath in dust (if she's a land fowl) or water (if she's a waterfowl) before zipping back onto the nest.

Try Not to Handle Her

As much as possible, avoid handling a broody hen. Too much attention can be off-putting enough to cause a hen to abandon her eggs. If you do need to handle a broody, first gently raise each wing. She may be holding an egg between a wing and her body; when you lift the hen, the egg may drop into the nest and break, possibly cracking other eggs as well.

FEEDING THE MAMA-TO-BE

If a hen doesn't seem to be getting off the nest to eat, encourage her either by lifting her off the nest and putting her down near the feed or by sprinkling a little scratch in front of the nest to pique her interest. If she remains indifferent, chances are she's been eating when you're not around.

One way to tell if a hen has been eating is to watch for broody poops. To avoid contaminating her eggs, a hen rarely poops in the nest. Instead, she holds her droppings until she leaves the nest, then immediately relieves herself by dropping one huge blob. Her poop will be less soupy — in or out of the nest — if you feed her scratch grain instead of layer ration. Scratch not only keeps droppings

more solid, and therefore easier to clean out, but also helps the hen maintain her weight.

A setting hen should start out at the peak of health with a good layer of body fat as indicated by a creamy or yellow hue to the skin; a hen with reddish- or bluish-looking skin has insufficient fat. Because she won't eat much while she's on the nest, she needs those fat reserves to see her through.

A hen on the nest will typically drink more than she eats. To encourage drinking and prevent dehydration, make sure her water is always fresh, clean, and plentiful.

PAPA'S ROLE

Laying and setting is solitary business for a chicken or turkey hen. Among ducks, however, a drake may stand guard while his hen lays an egg, but after she starts nesting he loses interest and wanders off on his own.

A guinea cock, too, will stand guard while his hen lays but will remain nearby — pacing up and down like the expectant father he is — while she broods. If he senses danger he may either sound an alarm or move off to draw the intruder's attention away from the nest. After the keets hatch, he

helps raise them and in fact may raise the keets alone should anything happen to their mother.

A gander becomes fiercely protective, standing guard and warding off all intruders, human or otherwise, throughout the brooding period. When the goslings arrive, the gander shares equal responsibility for protecting them from harm. So strong is a gander's parental instinct that, even without a mate, he may adopt an orphaned brood of goslings or ducklings.

The Hatch

After the sixteenth day for chickens and the twenty-fourth day for others, do not disturb the broody or her eggs. See that she has plenty to eat and drink, then contain your impatience until the little peepers pop into the world. While the birds are hatching, keep an eye on things but don't interfere unless your help is absolutely essential — such as to rescue early chicks wandering away and getting chilled while Mama stays on the nest to hatch the rest of the brood.

Normally all the babies will hatch within a few hours of each other. The hen will continue covering her brood on the nest for another day, maybe two, waiting for stragglers. Sometimes when a hatch is slow, the hen may hop off the nest to follow the earliest chicks, leaving the rest to chill and die while they're still wet or even before they fully get out of the shell. In such a case gather up and care for the first chicks, returning them to the hen after the hatch is complete.

Occasionally, a hen will be so horrified by the appearance of interlopers beneath her that she'll attack the little fuzzballs. Be ready to rescue the chicks and brood them yourself.

Typically a hen stays on the nest, covering her newly hatched brood for an extra day or two, waiting for stragglers.

More typically, a mother hen gets possessive and extremely aggressive against all intruders, including you. Any small children who might be excited to see the hatchlings should be kept at a distance to avoid being frightened or attacked by a belligerent hen. Even if the hen remains calm, children should be instructed not to pursue or handle the delicate hatchlings, and curious household pets must be kept away.

BROODING PEN

For the first few days, the hatchlings spend most of their time hidden under their mother. Even after leaving the nest, they remain close to the hen and periodically crowd under or around her for warmth. On cold or rainy days, she may open her wings like an umbrella to provide additional protection. In early spring when the weather is still cold and in rainy weather, confine the family to a shelter to ensure survival of the hatchlings.

Until the chicks are about 2 weeks old, they typically stick pretty close to the mother hen. At about 2 weeks they start getting adventuresome and wander off to feed on their own but frequently return to the hen for warmth, for comfort, and to sleep. As time passes, depending on their environment, the chicks may eventually go their separate ways or remain together as a social group.

While the babies are growing, a brooding pen offers the hen and chicks safety from the elements and from predators. A hen won't always find a safe place on her own. In my early years of keeping chickens, I let a broody make her nest in the layer house. When the chicks left the nest, they fell into the droppings pit and had to be rescued.

Later, another hen made her nest in the hayloft. One morning I found chicks wandering around the barn peeping miserably while the oblivious mother hen sat happily on her nest overhead in the hay. To prevent such mishaps, if you didn't move the hen to a safe brooding pen when she started setting, move her as soon as the hatch is complete. Moving a mother hen of any species is much easier than moving a setter; once the hatchlings appear she is quite unlikely to abandon them.

You might install her and her brood in the corner of a tool shed or other outbuilding. Or you could set up a dedicated brooding pen, which is simply a small shelter just big enough for a hen and her growing brood. No specific design is better than any other. You might adapt a doghouse, a cleaned storage drum, or an unused camper shell. If you make the shelter from scratch, it needn't be more than about 3 feet × 3 feet (1 m × 1 m) for chickens, guineas, or ducks; 4 feet × 4 feet (1.25 × 1.25 m) for turkeys or geese. Be sure the floor, walls, and roof are constructed tightly, without any holes, making them rat- and snakeproof, and install screened vents near the top for good airflow. Add a door that opens onto a small, enclosed run. If the setup is outdoors, be sure to provide shade.

Never let a hen and her brood just wander around an open yard, or the babies may fall prey to house cats, hawks, and other predators. Even if the yard is fenced for poultry, babies can slip through some mighty small openings. I once had some goslings commute through a chain-link fence to graze on the lawn; eventually, they got too fat to slip back into their own yard and had to be rescued from the maw of a neighborhood cat. Babies wandering loose might get chilled in damp grass or get lost in tall vegetation and not be able to find their way back. If one of her babies wanders out of earshot, a hen has no way of knowing the little one is missing. Confining the mother and her brood for at least a month will help prevent mishaps.

Confine the mother and her brood in a private predator-proof enclosure for at least a month.

Removing the Babies

If you're hatching particularly valuable poultry, you might wish to remove them from the hen's care and brood them yourself. Some breeders of rare or valuable birds jump the gun by moving term eggs to a mechanical hatcher. That's tricky business, though, since any delay while moving the eggs or an incorrect setting of the hatcher can prove disastrous. On the other hand, a properly functioning hatcher may be a safer option than leaving eggs to hatch under a hen that doesn't have enough of a track record for you to be sure of her competence as a mother. A hen thus deprived of her chicks may spend a day or two looking for them but will soon return to business as usual.

FEEDING THE BROOD

Feed the hen and her brood starter ration, as described in chapter 3. Because the hen won't resume laying while she's brooding, feeding her starter instead of lay ration won't hurt her, but the high calcium content of layer ration can harm the babies.

Initially, chicks may be fed in a large, open pan they can walk and scratch in. When the babies get big enough to eat from a trough, fasten the bottom to a short piece of 2×6 lumber to give it a sturdy base so the hen can't tip it over.

You'll also need a water container the hen can't knock over. A 1-gallon (4 L) waterer is ideal. If you're brooding bantams, for the first few days put marbles or pieces of clean gravel in the rim to prevent a tiny chick from climbing into the drinker and getting a chill or drowning.

Foster Moms: Not Too Picky

A broody chicken will mother any hatchlings that appear under her, but may become disconcerted if her brood jumps in for a swim. On the other hand, she is just as likely to wade right in and swim with them!

A setting hen doesn't know whether or not the eggs under her are her own, and even under natural circumstances the eggs in her nest may be a collection of eggs from various hens in the flock. So you might deliberately choose to have your broody hen hatch eggs laid by other hens, even of another species. I frequently slip guinea or turkey eggs under a chicken hen, and once I asked a chicken to hatch duck eggs, although she went into a tizzy when her kiddies jumped into the duck pond.

When asking a chicken to hatch eggs that are much larger than her own, such as turkey or goose eggs, she likely will have trouble turning them adequately. To ensure a successful hatch, pitch in and turn them for her. Mark them as you would for artificial incubation described in Manual Turning on page 167 in chapter 9, so you can be sure each egg has been properly turned.

If you want to combine eggs that differ in their incubation period, you'll need to do a little advance scheduling based on the incubation times for each species. You might, for instance, add chicken eggs to a nest full of guinea eggs after a hen has been on them for 7 days. Or you might start them all together and be prepared to gather the chicks into a brooder while the hen finishes hatching the guineas. In the latter case, she is unlikely to take back the chicks after a week, so be prepared to raise them yourself.

Sometimes, a chicken is not willing to continue brooding for the extra days needed to hatch eggs of other species. But most hens will successfully wait it out. Some, in fact, will brood far beyond any normal incubation period if their eggs don't hatch because they got too old, overheated, or chilled or were not fertile. In such a case you'll have to step in and break up these hens (see Discouraging Broodiness, page 127).

ENSURING CHICK ACCEPTANCE

A hen may be enticed to raise chicks she didn't hatch herself. She is most likely to accept foster chicks if she's been seriously setting for several days. The hatchlings must be not much more than a day old and still receptive to accepting (or imprinting) a new mom. To increase the chances both parties will accept each other, slip the babies under the hen at night. In the event things don't work out as you had planned, be prepared to gather them up and raise them yourself.

You can sometimes induce a hen to raise a brood even if she hasn't been setting. Put the hen and baby chicks together in a dimly lit brooding pen, and watch for the four typical signs the hen is willing to care for the chicks. Not necessarily in this order, a hen willing to accept chicks will:

- Spread her wings to cover the chicks
- Tidbit — sound the food call and pick up and drop bits of food
- Rush to the assistance of any chick making sounds of distress, such as if you pick it up
- Cluck loudly and continuously

A mother hen will take her brood under her wing, even if they are another species that grow to be larger than she is.

Since a hen can successfully mother as many as three times the number of chicks she hatches, you might artificially incubate additional eggs, with the intent of having your hen brood the resulting chicks along with her own. This ploy works best if all the chicks hatch at the same time. As always, slip the freshly hatched incubated chicks under the hen at night. If you're moving the hen and her brood from the nesting site to a brooding pen, place all the chicks in the pen first so they'll get mixed together before you introduce the hen.

PRECOCIAL PEEPERS

Under normal conditions a hen and her brood make initial contact through sound rather than sight, so a foster broody is more likely to accept chicks slipped under her at night and have the entire night to listen to them peep. Since each poultry species has its own distinct sounds, how do you suppose keets or poults, for example, that hatch under a chicken, so readily recognize her as Mom?

Any animal that can feed itself almost from the moment of birth is considered precocial, which certainly describes barnyard hatchlings that start pecking the ground soon after they hatch. The word "precocial" comes from the Latin word *praecox*, meaning mature before its time. The chief characteristic of precocial birds is their spryness soon after hatching, and as a result they may easily get separated from their mother. Once they hatch they don't have much time to learn to recognize the sound of her call, which is essential to their survival.

So an important characteristic of precocial birds is that they communicate with their setting-hen mama before they hatch. The embryos inside their shells peep and the broody hen responds. Thus the babies learn to recognize their mother's voice while they are still in the shell. Entering the world with the ability to quickly find their way back to Mom, the precocial peepers are ready to hit the ground running.

Precocial hatchlings learn to recognize their mother's voice before they hatch, even when they are not the same species as Mom.

Discouraging Broodiness

A lot of chicken keepers are thrilled to death when a hen goes broody. Others don't want their hens to brood because a brooding hen stops laying. If you keep hens primarily for table eggs, or you raise a rare or valuable breed and prefer to hatch as many of their eggs as possible in an incubator or under other hens, you may wish to discourage broodiness to keep your hens laying.

Hens, like people, don't always react as you expect them to, and a persistent broody may continue no matter what you do. Depending on how serious the hen is about setting, you might successfully discourage her, or break her up, using these techniques:

- Avoid letting eggs accumulate in the nest
- Repeatedly remove the hen from her nest
- Move or cover her chosen nesting site so she can't return to it
- Confine the hen to different housing
- Put the hen in a broody coop

The function of a broody coop is exactly the opposite of what its name might imply. It consists of a cage with an open wire or slat floor and raised off the ground. A cage that's hung so it sways when the hen moves works best. The hen is housed in the broody coop for as long as necessary to break her up, usually 1 to 3 days.

The longer the hen has acted broody, the longer she'll take to start laying again. A hen that's broken up after the first day of brooding should begin laying in 7 days; a hen that isn't broken up until the fourth day may not start laying for about 18 days.

Spending 1 to 3 days in a hanging cage with a wire or slat floor is usually enough to discourage a hen from brooding.

7

SELECTING AN INCUBATOR

Incubators come in a broad range of sizes, styles, and prices. The smallest ones will fit on a bookshelf. The large cabinet models take up a considerable amount of floor space. Between the two extremes is an array of tabletop models in an assortment of sizes.

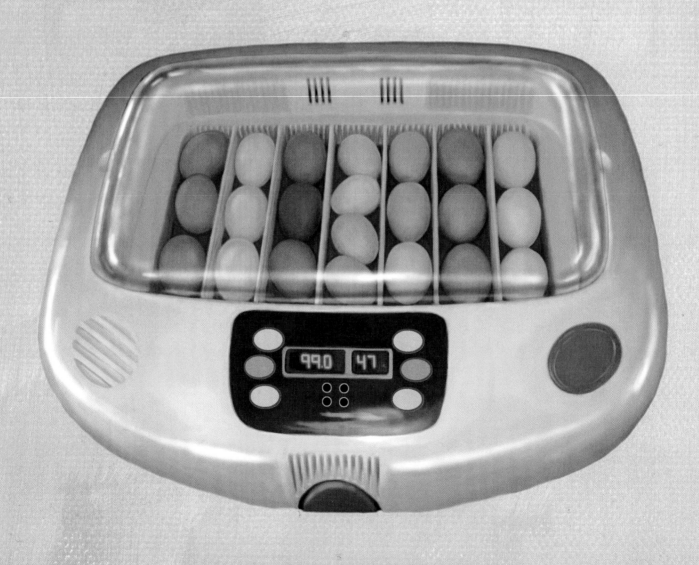

Egg Capacity

Incubators for home use hold anywhere from 3 eggs to 300. In determining what size you need, remember that not all the eggs you put into the incubator will hatch; a reasonable hatch rate is considered to be 80 to 85 percent. If you have a 48-egg incubator — and assuming all the eggs you put into it are fertile and everything goes well — you can expect about 40 hatchlings. With experience you may coax your incubator to a higher hatch rate; with less attention to detail, you'll get a lower rate. A fully automated computerized digital incubator can approach a hatch rate of 100 percent.

Some incubator manufacturers specify how many duck or goose eggs their units will hold. Others mention only chicken eggs. When only chicken eggs are specified, the incubator may not be big enough to handle the larger eggs laid by turkeys or waterfowl.

Most of the larger models can handle any kind of egg. To estimate the equivalent number of waterfowl eggs, figure 75 percent of the chicken egg capacity for duck eggs, 60 percent for Muscovy eggs, and 40 to 30 percent for turkey or goose eggs (depending on their size). That 48-egg

This digital floor model incubator by GQF has a capacity of 270 chicken eggs or the equivalent.

Features to Consider

Among all the various sizes, available features range from minimal — basically, a heated box — to fully automated and digitally controlled. When deciding which incubator is right for you, consider these features:

- Capacity: how many eggs it holds
- Turning: manual or automatic
- Airflow: still air or forced air
- Temperature control: mechanical or electronic
- Humidity control: manual or automatic
- Ease of observation
- Ease of cleaning
- Price

A low-cost incubator generally has the fewest features, lacks a turning device, and may less accurately control temperature and humidity. Such an incubator, compared to a more expensive full-feature model, requires closer supervision to get good results. So in considering price, weigh the time factor — do you want to spend a lot of time fussing over your incubator and will you always be readily available to do it?

incubator theoretically could handle 36 duck eggs, 29 Muscovy eggs, and about 16 goose eggs.

The smallest incubators are designed for educational purposes, to show classroom or home-schooled students how a chick develops from an egg. The large floor-model cabinet incubators are intended for serious breeders who hatch hundreds of eggs each season. In between are plenty of size options suitable for the average backyard small flock owner.

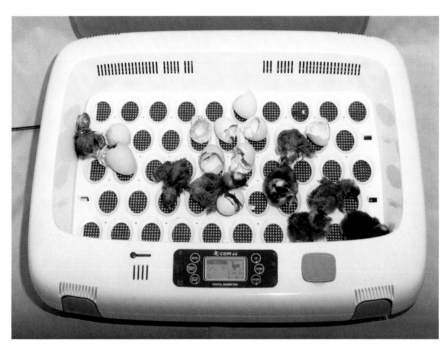

This preprogrammed and fully automated digital incubator by R-Com has a capacity of 50 chicken eggs.

Turning Devices

By fidgeting in her nest and adjusting eggs with her beak or bill, a setting hen periodically turns each egg beneath her. In doing so, she keeps the yolk centered within the white. The yolk inside an egg that isn't turned will eventually float away from the center and stick to the shell membrane, and the developing embryo will die.

To imitate the hen's activities, some incubators have automatic motor-driven turning devices that either roll the eggs or tilt them from side to side at regular intervals. The turning device might be a rotating ring or plate, a tray that rocks from side to side, or a series of racks that individually tilt back and forth.

In most tilting incubators the eggs are held 45 degrees from vertical, alternately in opposite directions. However, hatchability is better in incubators that tilt eggs at an angle of 75 degrees, even 90. In the latter case, the eggs lie nearly horizontal in one direction, then travel in a 180-degree arc to lie horizontally in the other direction.

The rotating device or tilting trays must be correctly sized according to the type of eggs you plan to hatch. The incubator may come equipped for chickens, with other devices available as extra options. Or in some cases the incubation unit and turning devices might be sold separately so you can select exactly the size you need.

The cabinet incubator I use now has egg trays that tilt at an angle of about 45 degrees and periodically rotate from one side to the other. Before I got this incubator I used a couple of smaller tabletop models with racks that rotated individually. As the racks tilted back and forth, some of the eggs came into contact with mechanisms attached to the cover and got crushed. Using an indelible pen I marked the positions of those trouble spots in the racks so I would remember not to put eggs there in the future. An alternative would have been to purchase an expansion ring to place between the base and lid, giving the eggs a little more room.

Some incubators either aren't equipped with a turning device or else offer the turner as an optional feature. Price may be a consideration in deciding whether to get an incubator with or without automatic turning, but your time may be of equal concern. Without a turner, you'll need to be on hand at least three times a day, every day, to turn the eggs yourself. The procedure is described on page 167.

Airflow

Developing embryos use up oxygen rather rapidly, while at the same time generating carbon dioxide. An incubator therefore needs a good airflow to constantly replenish oxygen and remove the carbon dioxide. All incubators have vents to admit fresh air and expel the stale air. Some small incubators rely on gravity to let warm, humid air escape through vents in the cover, which in turn causes fresh air to be drawn in through vents in the bottom. Some small units and all large units have a built-in fan to circulate the air.

A unit without a fan is called a **still-air incubator** but may also be described as a gravity-flow, gravity-ventilated, or natural-draft incubator. A unit with a fan is a forced-air incubator, also designated as a circulated-air or fan-ventilated incubator.

Although still-air incubators are less expensive, a **forced-air incubator** steadily circulates warm air to maintain a more uniform temperature

How Gravity-Flow Incubators Work

A still-air incubator circulates air with the help of gravity, and therefore is often called a *gravity-flow incubator*. Both warm and cool air are subject to the forces of gravitational pull. But cool air is denser than warm air and therefore subject to a greater pull by gravity. As gravity pulls cool air below the level of less dense, lighter, warmer air, it displaces an equal amount of warm air, forcing the warm air upward (in this case, through the vents at the top of the incubator). So what makes a gravity-flow incubator work is that cold air sinks, which doesn't sound nearly as catchy as saying "hot air rises."

Talking Fowl

cabinet incubator. A large-capacity incubator designed to be placed on the floor

egg turner. A device that either tilts or rolls eggs to periodically change their position during incubation

forced-air incubator. Also called *circulate-air incubator* or *fan-ventilated incubator*; an incubator with a built-in fan that constantly circulates air to maintain an adequate oxygen level and keep the temperature even throughout

hatcher. An incubator that lacks a turning mechanism because it is used only during the hatching phase of incubation

hygrometer. An instrument for measuring humidity in the air

humidity pads. Sponges used in an incubator to increase the surface area available for evaporation, thus increasing the humidity level; also called *wick pads*

still-air incubator. A mechanical device for hatching eggs that lacks a fan to circulate air; also called *gravity-flow incubator* or *gravity-ventilated incubator*

tabletop incubator. Small-capacity incubator that usually holds 50 eggs or fewer and is typically placed on a table or countertop

vent holes. Small holes cut into the top or sides of an incubator for ventilation to release stale air and admit fresh air throughout the incubator, resulting in a better hatch rate. Additionally, in an incubator lacking a fan, humidity can't be accurately measured.

In tabletop models the fan is usually in the cover. When you remove the cover, you move the fan away from the eggs. A floor model usually has the fan at the top or back of the cabinet and should have a switch that lets you turn off the fan before you open the door, to avoid blasting cool air across the warm eggs. A unit that lacks such a switch must be powered down before the door is opened.

My first cabinet incubator, a vintage unit made of solid redwood, had separate off switches for the fan and heater. When I needed to open the door while the incubator was in use, I flipped off the fan so it wouldn't draw in cool air while the door was open. My newer, modern incubator has no separate fan switch. In fact, it has no off-switch at all. The only way to turn it off is to pull the plug from the wall outlet. Worried about wear and tear to the plug from frequent yanking, I asked my husband to change the wall outlet to a switched outlet. Now when I want to turn off the incubator I just flip the wall switch.

Turning trays, such as those in this GQF cabinet incubator, automatically tilt alternately to the left and to the right.

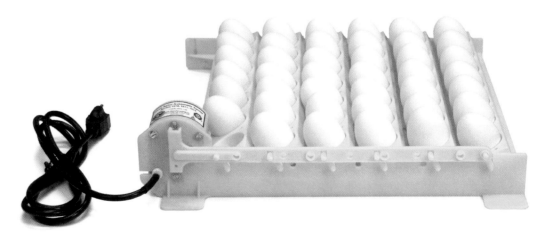

These turning racks by GQF are equipped with a little motor that automatically tilts each row from side to side.

Temperature Control

The normal body temperature of a hen of any poultry species is about 103°F (39.5°C). You'd think that would be the temperature at which an incubator should be run, but other factors figure in. For one thing, the hen's body does not surround the eggs beneath her but rests only against the tops, which results in a slightly cooler temperature within the eggs. Furthermore, an incubator's thermometer is not designed to read the temperature inside the eggs, and is not necessarily positioned to read the temperature at the position of the eggs. As a result, operating temperatures recommended by incubator manufacturers vary slightly from one model to the next.

THERMOMETER

A fully electronic incubator shows the temperature readout as part of its digital display. Others use a thermometer, which may itself be electronic. Make sure your thermometer is true by checking it against a second or even a third thermometer. Put them together into a cup of water at about 100°F (38°C). If two agree, they are probably both accurate. If two don't agree, add a third. Most likely, two will agree; the third one, the one that's different from the other two, is likely the one that's off.

Some thermometers are designed for placement inside the incubator at the level of the eggs but are so small you can't easily read them without opening the incubator, which of course causes the temperature to drop before you get a reading. Either a stem or a long probe thermometer solves that problem; it has a probe you insert into a hole in the incubator while the readout dial remains on the outside.

An electronic incubator, like Brinsea's Mini Advance, displays the temperature and other important information in digital format.

This long-probe digital thermometer registers percent relative humidity as well as temperature.

A stem thermometer has a probe you insert into a hole in the incubator while the readout dial remains on the outside.

THERMOSTAT

An incubator with an electronic thermostat is preset by the manufacturer and should need little or no tweaking for the first few years. Some digital incubators include a readout that tells you if the room temperature outside the incubator is too hot or cold for the incubator to maintain proper internal temperature. Regulating the heat in an electronic incubator is done with convenient touch pads or touch buttons.

Even a high-tech electronic incubator may have as a backup an old-fashioned wafer thermostat, which is a metal disk filled with ether. When the incubator heats up, the ether expands, causing the wafer to swell until it makes contact with a button switch that turns the heat off. As the incubator cools down, the ether contracts until the wafer loses contact with the switch, causing the heat to go on again.

Regulating a wafer involves turning a control bolt to adjust the temperature at which the wafer makes contact with the switch. Most wafer-controlled incubators have an indicator light that goes on when the heat is on and off when the heat is off. After a while you develop a sixth sense that makes you feel uneasy if the light stays on or off for too long.

Unlike an electronic thermostat, a wafer is affected by barometric pressure, so weather fluctuations can cause temperature swings. Additionally, a wafer typically takes longer than an electronic thermostat to bring the temperature back up after the incubator has been opened. It also produces a less steady temperature; the cycling of the wafer against the thermostatic switch results in greater fluctuations — as much as 2° above or below the desired temperature.

When a wafer ceases to function properly because it springs a leak, causing the ether to escape, the switch remains in the on position, and the temperature stays too high for too long. Unless you catch it right away, your eggs will cook. If your incubator uses a wafer, always keep at least one spare wafer on hand.

TEMPERATURE FLUCTUATIONS

A failure of the thermostat, whether electronic or wafer controlled, is not the only thing that can cause the incubator's temperature to soar. The incubator could be too close to a heater, for example, or sunlight could fall on the incubator during part of the day. Even just a short-term rise in temperature does greater damage than a temperature drop, which may occur if the incubator's cover or door is left ajar, hatching fluff jams the wafer switch, a child or pet pulls the plug, or the power goes out.

The more often you check the temperature, the more likely you are to catch and correct problems, such as a pulled plug or a wafer gone bad. During my early years using an incubator, I kept a piece of paper nearby on which I recorded the time and

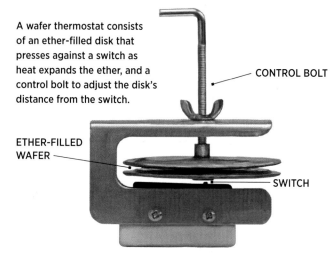

A wafer thermostat consists of an ether-filled disk that presses against a switch as heat expands the ether, and a control bolt to adjust the disk's distance from the switch.

CONTROL BOLT

ETHER-FILLED WAFER

SWITCH

Wafer Check

A temperature that fluctuates more than usual may be an indication the wafer is on its way out. To test a wafer for soundness, hold it briefly under a faucet while running warm water.

- A good wafer expands quickly
- A failing wafer expands slowly
- A bad wafer doesn't expand; toss it

temperature every time I passed the incubator. After a while checking got to be a habit, and I no longer needed the paper reminder. But a paper record of temperature checks is a good idea when two or more people share the responsibility.

The better the incubator is insulated, the less it is affected by fluctuations in room temperature, such as might be caused by sun coming through a window, heat or air conditioning turned off during the night, or heavy storms that cause drastic changes in barometric pressure. Occasional minor fluctuations are normal, so once the incubator's temperature has been set, stop fiddling with it.

Some cabinet incubators have an alarm that sounds or lights up when the temperature drops or rises past a certain range. If you're handy with gadgets, you could easily rig up something similar for an incubator lacking such an alarm.

Humidity Control

For a successful hatch, moisture must evaporate from the eggs at just the right rate. Overly rapid evaporation can inhibit the chicks' ability to get out of their shells at hatching time. Overly slow evaporation can lead to mushy chick disease (omphalitis), in which the yolk sac isn't completely absorbed so the navel can't heal properly; as a result bacteria invade through the navel, causing chicks to die at hatching time and for up to 2 weeks afterward. Your incubator's operation manual will indicate the correct moisture needed throughout incubation and at the time of hatch.

PROVIDING MOISTURE TO THE EGGS

Evaporation is regulated by the amount of moisture in the air — the more moisture-laden the air, the more slowly moisture evaporates from the eggs and vice versa. To reduce the rate of evaporation from eggs, every incubator has a water container of one sort or another that gradually releases moisture into the air. Some devices must be filled manually. Others may be fitted with an external water container that automatically feeds into the incubator, which must be checked occasionally and refilled as needed so it never runs dry.

The water-holding device might be a simple pan, a divided pan, or grooves molded into the bottom of the incubator. Adjusting the surface area available for evaporation regulates humidity. Using a pan with a larger surface area increases humidity; a smaller pan decreases it. Filling more divisions or grooves with water increases humidity, and filling fewer decreases it. Increasing surface area with sponges sold as humidity pads or wick pads increases humidity; partially covering the water pan with foil decreases humidity. A fully automated incubator has a pump that adjusts humidity without human intervention, aside from filling the reservoir.

Incubation humidity may be fine-tuned by adjusting the incubator's vents. Some vents must remain open at all times for good oxygen flow; others have either removable plugs or sliding covers that allow you to adjust the size of the openings. Closing vents increases humidity by trapping more moisture-laden air within the incubator. Opening vents decreases humidity by allowing more moisture-laden air to escape.

To ensure the incubator never runs dry, this 18-egg incubator by Lyon Technologies has an external water container that lets you easily see when it needs refilling.

MEASURING HUMIDITY

An all-electronic incubator uses computer technology to sample the humidity as well as the temperature of the air within the incubator and gives you a digital readout as percent relative humidity. The same technology is used by an electronic hygrometer — a battery-operated device you can place in your incubator to take humidity measurements.

In a nonelectronic incubator, humidity may be measured by a hygrometer devised by fitting a regular thermometer with a cotton wick or "sock" moistened with warm water. As the water evaporates from the wick, the thermometer will give you a lower reading than it would without the wick. The converted thermometer thus measures humidity in terms of wet-bulb degrees, in contrast to the dry-bulb degrees by which the thermometer normally measures heat.

The fitted thermometer must be placed in your incubator where you can see it when the cover is on or the door is shut. After a few minutes the wet-bulb temperature will stabilize, and you can take a humidity reading. A wet-bulb reading will be more accurate in a forced-air incubator than in a still-air incubator.

To keep running tabs on humidity, use a longer wick and dangle the tail end into water, which might be the incubator's humidity pan or a separate container of water placed near the thermometer. When the latter is filled with distilled water, the wick lasts longer because it is less likely to get crusted with mineral solids, especially if your tap water tends to be hard.

You will need spare wicks to replace those that lose absorbency over time. You can buy wicks from some poultry supply outlets, or you can make them from cheesecloth or gauze bandages. I have used a 4-inch gauze pad, opened out and refolded twice the long way, then stitched up one side to make a 1-inch tube, which I then cut in half, getting two wicks from each pad.

Ease of Observation

If you want to watch what's going on inside your incubator, especially during the hatch, opt for a model with a large observation window, clear plastic cover, or transparent door. Incubators designed for classroom use typically have large observation windows. Many incubators have tiny little windows, or none at all, which can be a problem for someone who is curious enough to keep opening the incubator to see what's happening. Doing so at hatch time can disrupt the hatch by reducing the humidity or the temperature, or both.

Sometimes the incubator's interior design can defeat the purpose of its observation function. I once had a tabletop incubator with a clear plastic cover, but all the heat coils, fans, and other gizmos attached to the cover made it impossible to see anything inside, including the thermometer.

An incubator with a small window, or no window, must have an external means of monitoring heat and humidity. An electronic incubator with a digital display has this covered. A nondigital incubator should be fitted with at least a stem thermometer so you don't have to open the incubator to get a reading, which of course won't be accurate no matter how quick you are.

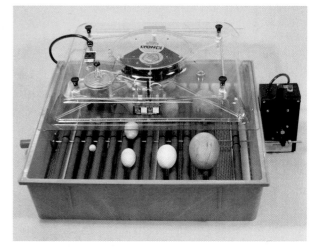

Lyon's Roll-X incubator features a transparent cover for easy observation of eggs during incubation and hatching

Ease of Cleaning

Hatching makes a mess. There's no getting around that. Gunk accumulates in the bottom of the incubator, and the heater and other controls get coated with a layer of fluff that can eventually gum up the works. So you'll be happy if your incubator is easy to clean and sanitize.

Most tabletop models have a top and bottom that may be separated for a good scrubbing. But if the incubator is made of Styrofoam, you'll have to be careful not to crack or break the Styrofoam, and you certainly can't take a stiff brush to it. Because cleaning Styrofoam is such a problem, some manufacturers offer the option of a plastic liner that may be removed for cleaning, thus protecting the bottom half of the incubator from cleanup wear and tear. Hard plastic, fiberglass, and wood are more conducive to scrubbing.

A cabinet incubator can present a different set of problems. My cabinet incubator is so deep I can't reach all the way to the back to wipe the shelves. Also, the controls are at the back, behind a protective screen, and cleaning them requires unscrewing and removing the back panel. Since my incubator is in continuous use throughout the hatching season, I can't just shut it down and disassemble it for cleaning without jeopardizing the partially incubated eggs. It's otherwise an excellent incubator, and the most popular cabinet model on the market, so for quick and easy cleaning in midseason, as well as a thorough cleaning at season's end, my husband hinged and latched the back the same way the front door is hinged and latched.

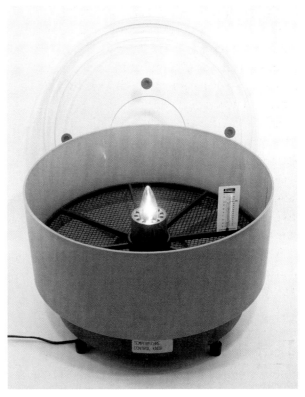

The Top Hatch by Brower can be tricky to operate as an incubator, but with its transparent cover, internal light, and ease of cleaning it makes a handy hatcher.

Incubator versus Hatcher

How much hatching you plan to do and how many different breeds or species you raise will determine whether you opt for one or more small incubators for single-stage incubation, one large incubator for continuous hatching, or a hatching unit that's operated separately from the incubator. Each option has advantages.

SINGLE-STAGE INCUBATION

Single-stage incubation, or all-in all-out incubation, means all the eggs in the incubator are scheduled to hatch at the same time. To achieve this outcome, a tabletop incubator that doesn't hold many eggs to start with is used. Then after the eggs are fully hatched you can clean out the incubator thoroughly in preparation for the next setting. Another advantage is that you can optimize your success rate by adjusting the temperature, humidity, and ventilation as incubation progresses to the hatching stage.

Single-stage incubation is the procedure to follow if you wish to hatch a few birds once a year. They might be all chicken eggs, or you might mix species by staggering your schedule — for instance, you'll need to put guinea eggs into the incubator one week before adding chicken eggs, since the incubation period for guinea eggs is about a week longer than for chickens. After the hatch, clean and sanitize the incubator and store it away until next season.

Another way to use single-stage incubation is to hatch one breed or mating at a time and brood them separately or hatch one species at a time to accommodate their differing incubation periods

Digital Recalibration

Electronic incubators require periodic recalibration, a process typically described in the user's manual. Recalibration requires the use of a separate and accurate thermometer and humidity meter. An incubator dealer with good customer service should be able help you recalibrate your digital incubator.

and hatching requirements. You would incubate, for instance, a batch of duck eggs and, after they hatch, clean out the incubator and start a batch of chicken eggs. You might even run more than one incubator at a time — one for each species.

CONTINUOUS HATCHING

Continuous hatching, or multi-stage incubation, means groups of eggs are placed in the incubator at different times and therefore hatch at different times. It is the typical backyard scenario for larger cabinet incubators. A common cabinet model has three incubation trays and, below them, a single hatching tray. This setup was initially designed for hatching chicken eggs because it allows you to fill one tray each week and, starting in 21 days, have chicks hatch once a week on the hatching tray for as long as you keep refilling the empty incubation trays. An advantage is that you can hatch more chicks and on a regular basis.

Continuous hatching can result in difficulty in keeping track of which eggs are about to hatch. To avoid confusion, you might label the incubation trays as A, B, and C. I choose letters instead of numbers to avoid including more numbers in my hatching records than are already furnished by setting dates, setting times, quantity of eggs set, and hatching dates. The tray labels are handy for hatching chicken eggs but are indispensable when I include turkey or guinea eggs, which take a week longer to hatch. By including the tray letters

on my hatching chart, I no longer have trouble keeping track of which eggs in which trays will hatch when.

Continuous hatching is possible (though not practical) with a tabletop incubator in which eggs must be turned manually. It is not a handy option for one with automatic turning, especially where the turner must be removed as hatching time approaches for some but not all of the eggs. Another problem is that eggs due to hatch require higher humidity than those under incubation. If the humidity is adjusted for the hatch, it is too high for the eggs still under incubation. If it is adjusted for incubation, it will be too low for a good hatch.

Yet another disadvantage of hatching continuously in any incubator is that hatching debris builds up and must be dealt with while some eggs are still under incubation. Despite your best efforts, you can't thoroughly sanitize the incubator with eggs in it, and as a result the hatching rate gradually declines. For this reason, if you plan to hatch lots of eggs, you might consider getting a separate hatcher.

USING A SEPARATE HATCHER

A separate hatcher is a unit into which eggs are moved from the incubator a couple of days before they start hatching. After the hatch, the hatcher is shut down, cleaned and sanitized, and made ready for the next hatch. Since hatching temperature and humidity are slightly different from incubation temperature and humidity, using a hatcher improves the hatch rate compared to continuous hatching.

A hatcher is identical to an incubator but with no egg turner. Dedicated high-tech hatchers are available for sale, but an inexpensive still-air incubator makes a less expensive option. Using a separate hatcher offers several important advantages.

Optimal hatching environment. The hatch rate can be improved over continuous hatching, because the incubator may be run at the optimal temperature and humidity for incubation while the hatcher is set for optimal hatching temperature and humidity, as described in chapter 9.

Cleanliness. The incubator remains more sanitary because hatching debris — broken shells, chick fluff, and embryo wastes — is confined to the hatcher, which after each hatch may be thoroughly cleaned and sanitized before the next hatch is moved from the incubator.

Mother's helper. The hatcher may be used in combination with a setting hen, where predators or aggressive mature poultry might endanger hatchlings. To ensure their survival, move naturally incubated eggs to the hatcher when they are ready to hatch, then brood them indoors.

When selecting a hatcher, make sure its capacity matches the number of eggs you plan to hatch at a time. For instance, if you get a three-tray cabinet incubator that holds eight dozen chicken eggs on a tray, and you fill the trays weekly, you would need a hatcher that holds at least 85 chicken eggs, allowing for normal attrition of about 10 percent; see "Candling the Eggs" on page 172.

Be Creative: Make Your Own Incubator

If you like do-it-yourself projects, you might consider making your own incubator, now that you know what the essential features are. A keyword search for "homemade incubator" will yield a wide variety of plans from as simple as adapting a cardboard box to as elaborate as any fancy woodworking project. Many of the plans show how to reduce costs by upcycling such items as a Styrofoam ice chest, a picnic cooler, a glass aquarium, a discarded cabinet or bureau, a gutted refrigerator or freezer, or a defunct wine cooler. When you find a plan you like, check the fine print to ascertain whether or not the designer successfully hatched eggs in it.

I have had people tell me they hatched eggs in an electric frying pan, although I don't recommend it because controlling the heat and especially the humidity is tricky business. More successful has been using an electric pan as a hatcher to provide gentle heat to eggs that are about to hatch or are already in the process of hatching.

8

EGGS FOR HATCHING

Hatching eggs are fertilized eggs that will hatch if properly incubated. They may come from your own flock or be acquired from an outside source. Either way, you'll have greatest success by taking care to select the best eggs for hatching and storing them under optimal conditions until they are put under a setting hen or into an incubator. Understanding a little bit about egg structure will help you in your quest to select ideal eggs for hatching.

What's in an Egg?

An egg is composed of four fundamental components: the yolk with its associated membranes, the germinal disc or blastodisc, the albumen or egg white, and the shell with its associated membranes. When an egg is incubated, each of these four components plays an important role in the development of the embryo.

THE YOLK

The yolk accounts for about one-third of an egg's weight and contains all of the egg's vitamins, along with trace amounts of various minerals. It consists primarily of a yellow fat- and protein-filled fluid in which an embryo forms and from which the embryo obtains nutrients.

The yellow color of a yolk is due to xanthophyll, a natural fat-soluble yellow-orange pigment found in green plants and yellow corn. A hen that eats green plants, for example, produces eggs with dark orangey yolks. A hen that eats yellow corn or alfalfa meal produces yellow yolks. The yolks from a hen fed white corn may have almost no color at all.

Carotenoids, the plants containing xanthophyll, are important because they are a source of vitamin A. This vitamin is among the antioxidants in the yolk that protect a developing embryo from damage caused by free radicals, which are normal but potentially harmful by-products of an embryo's metabolism. Vitamin A also helps preserve the maternal antibodies intended to immunize the chick from diseases in the hen's environment.

The yolk is not of uniform color throughout but consists of concentric rings of yolk enclosed within four membranes that separate the yolk from the albumen. At the center of the yolk is a ball of white yolk, surrounded by alternating layers of thick dark yolk and thinner white yolk. Although you might never see it — except maybe in a hard-cooked egg — a neck of white yolk extends from the center to the edge of the yolk, flaring out and ending just beneath the blastodisc.

YOLK STRUCTURE

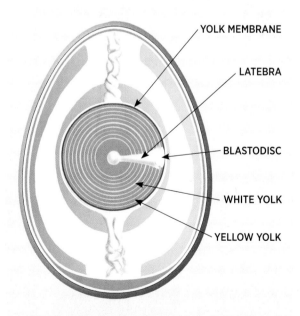

The yolk consists of concentric white and yellow rings enclosed within membranes that separate the yolk from the albumen. At the center of the yolk is a ball of white yolk with a neck that flares out (collectively, the latebra) and ends just beneath the blastodisc.

Talking Fowl

fertilized egg. An egg containing sperm and therefore capable of developing into a baby bird

hatch. The process by which a newly developed bird escapes from within the eggshell

hatchability. The ability of fertile eggs to hatch, expressed as a percentage of those that hatch in a group placed together under incubation

hatching egg. A fertilized egg that has been properly handled to best maintain hatchability

THE BLASTODISC

On the surface of the yolk is a small circular white spot that is 2 to 3 millimeters in diameter and is usually visible on top of the yolk when you crack an egg into a bowl. This spot is where the sperm joins the egg. In an egg that has not been fertilized, this spot is known as the germinal disc or blastodisc. In a fertile egg it's called the blastoderm and contains all the genetic material contributed by both parents that is needed to produce a baby bird.

When an egg is incubated, the embryo develops from the blastoderm. As the embryo develops, it gradually sends a system of blood vessels into the membranes surrounding the yolk. These blood vessels carry nutrients from the yolk to the developing embryo.

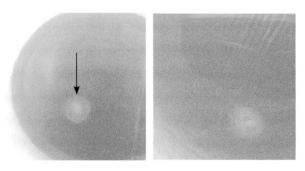

The blastodisc is a small irregularly circular white spot on the surface of the yolk that, when fertilized, becomes a blastoderm, consisting of a series of concentric rings.

THE ALBUMEN

The albumen, or egg white, is a clear fluid surrounding the yolk. It is 90 percent water combined with approximately 40 different proteins, one class of which consists of globulins (so called because of their globular shape) that confer immunity. The albumen is made up of four layers, two thick and two watery.

The inner thick, or chalaziferous, layer surrounds the yolk and lies against the outer yolk membrane. It cushions the yolk and contains defenses against bacteria. On two sides of the yolk, extensions of this layer are twisted together to form fibrous cords, or chalazae, from which this layer derives its name. The chalazae anchor the chalaziferous layer to the shell membrane at both the pointed and blunt ends of the egg. These anchors protect the yolk by suspending it and centering it within the albumen and also keep the blastodisc oriented upward.

The inner thin layer is a watery layer of albumen surrounding the inner thick layer and separating it from the outer thick layer.

The outer thick, or firm, layer of white is a gel that makes up the greater portion of the albumen.

ALBUMEN STRUCTURE

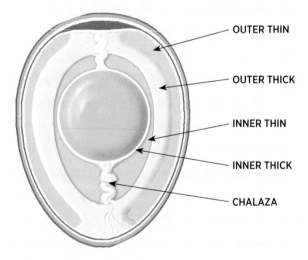

OUTER THIN

OUTER THICK

INNER THIN

INNER THICK

CHALAZA

The egg white consists of four distinct layers of alternating thick and thin albumen, the innermost of which anchors the yolk to the shell membrane by means of two cord-like chalazae.

The outer thin layer is a narrow layer of watery white lying next to the shell membrane. It repels bacteria by virtue because of its alkalinity and its lack of nutrients needed by bacteria for growth.

The albumen serves several important functions. It furnishes protein and water to the developing embryo. It acts as something of a shock absorber to protect the embryo from being jarred by sudden impacts. It shields the embryo from temperature changes. And it serves as a lubricant that helps the chick turn when it's ready to break out of the shell.

THE SHELL

The hard outer shell of an egg, which houses and protects the embryo, is lined with two membranes that control the evaporation of moisture from within the egg and protect the embryo from bacterial invasion. The inner membrane encloses the albumen and yolk. The outer membrane sticks to the inside surface of the shell.

After an egg is laid and begins to cool, the two membranes separate at the large end of the egg to form a pocket of air, or air cell. As an egg ages and moisture evaporates from it, the contents shrink and the size of the air cell increases. During incubation the air cell increases rapidly, and its size at various stages of incubation is an indication of whether humidity is optimal for embryo development and hatching.

The shell accounts for about 12 percent of an egg's weight. It is quite an elaborate structure consisting of these three basic layers:

The mammillary layer is the innermost layer, to which the outer shell membrane adheres. It serves as the foundation on which the rest of the shell is built.

The spongy, or calcareous, layer is the major portion of the shell. It is made up of tiny **calcite (calcium carbonite) crystals** — like little pencils standing on end — some of which fuse together, leaving pores between them that allow moisture and carbon dioxide to escape from the egg and oxygen to get inside to form the air cell. From this layer the developing embryo derives calcium for bone development.

The **bloom**, or **cuticle**, consists of dried mucus deposited by the hen just before she laid the egg. The cuticle seals the pores to minimize evaporation from the egg and prevent bacteria from entering through the shell. It is also responsible for the color of brown, tinted, and speckled eggs.

Generally, but not always, the larger the egg, the thicker the shell. The eggs of turkeys and geese are quite thick and tough. But so are guinea eggs, which are smaller than chicken or duck eggs. The thicker the shell, the more calcium is absorbed by the developing embryo and the stronger the skeleton of the hatchling. A strong skeleton is important for baby poultry that have to keep up with an active mama to survive.

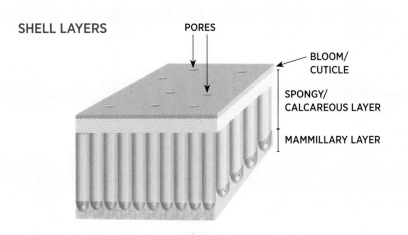

SHELL LAYERS   PORES

BLOOM/
CUTICLE

SPONGY/
CALCAREOUS LAYER

MAMMILLARY LAYER

Egg Selection

A breeder flock is strongest and healthiest in spring, making spring hatchlings the strongest and healthiest as well. Birds hatched in cool spring weather have time to gradually develop immunities before being exposed to large numbers of germs that proliferate with the approach of warm, humid summer weather. For general health and vigor, in most areas the best months to collect poultry eggs for hatching are February and March. In the far north, where the weather stays cold long into spring, March and April are better months for hatching. When selecting eggs for hatching, choose those that are of a normal size, shape, and color for your species and breed.

EGG SIZE

The size of an egg is influenced by the size of the yolk, which is relative to the size of the hen. Geese and turkeys lay larger eggs than chickens and ducks, which in turn lay larger eggs than guineas and bantams. Within each species big, healthy hens generally lay larger eggs than smaller, less vigorous hens.

Chickens lay larger eggs as they get older, although a really old hen may lay eggs that are smaller than usual. Pullets lay small eggs, and even after they start laying eggs of normal size, sometimes the last egg in a clutch will be small. Eggs that are smaller than normal generally produce smaller, less vigorous hatchlings. Excessively large eggs tend to hatch poorly, and those that do hatch may result in chicks that have unabsorbed yolks (soft bellies that don't heal) or inconsistent growth rates.

POULTRY EGGS

turkey · chicken · bantam · guinea · duck · goose

EGG SHAPE

An egg takes on its shape at about the same time as it acquires its shell, which occurs in a hen's shell gland or uterus. Therefore eggs laid by a particular hen are, with occasional exceptions, pretty much all the same shape. Since egg shape is hereditary, hens within the same family lay eggs that all have basically the same shape.

A healthy hen is more apt to lay eggs of normal shape than is an unhealthy hen. A healthy hen has strong uterine muscles with which to shape the egg. A less healthy hen or an aging hen with weakened muscles may lay eggs that are more round than oval. They may be blunt at both ends, making them almost completely round, or they may be slightly pointed at both ends.

Age influences egg shape. A pullet laying small eggs generally produces rounder eggs, while the eggs of an older hen tend to be more elongated. Wrinkled or misshapen eggs are often produced by pullets that aren't yet fully geared up for egg production and by tired, aging hens that are winding down. Stress, either mental or physical, can also result in wrinkled or misshapen eggs.

Season exerts influence on egg shapes. Eggs laid in the summer and fall are generally more elongated than those laid in winter and spring. Seasonal layers therefore produce more uniformly shaped eggs than poultry that lay year-round or nearly so.

Ideal Hatching Eggs

The ideal hatching egg has these qualities:
- It is of typical shape, size, and color for the breed
- The shell is clean and unstained
- The shell is smooth (not rough)
- The shell is sound, lacking cracks or thin spots

IMPERFECT EGGS

Too round

Too long and pointy

Too oblong (with a blunt end at both ends, making it difficult to determine which end is truly the blunt end)

Wrinkled

Misshapen

To ensure that the birds you hatch have the best start in life, select eggs that are most likely to come from normal, healthy hens. Don't hatch any that are round, oblong, or otherwise oddly shaped, or those with shells that are wrinkled, glassy, or abnormal in any other way. Such eggs may, of course, be reserved for eating.

SHELL COLOR

Shell color is a genetic trait that varies with species and breed. Turkeys and guineas lay pale eggs with speckled shells. Ducks lay eggs with white or pale green shells. Geese lay white-shell eggs.

The shells of eggs other than white are colored by pigments added to the shell toward the end of an egg's formation. Among chickens, brown-egg layers produce eggs of varying shades ranging from barely tinted to nearly black, influenced by more than a dozen different genes that affect shell color. The darkest shells come from Barnevelders, Marans, Penedesencas, and Welsumers. Most of the pigment of a brown-shelled egg is deposited in the bloom, leaving the inside of the shell pale or nearly white.

Ameraucanas and Araucanas lay eggs with blue shells, and the blue pigment is spread throughout the shell, which is therefore just as blue on the inside as on the outside. Green shells come from crossing a blue-egg layer with a brown-egg layer, resulting in a blue shell with a brown coating. The many different shades laid by so-called Easter Egg chickens result largely from blue shells coated with different shades of brown bloom.

The blue-egg gene is dominant, so a hen that lays blue eggs is not necessarily a purebred Ameraucana or Araucana. Conversely, an Ameraucana or Araucana hen that lays nonblue eggs is not purebred. Ameraucanas or Araucanas that lay eggs with green shells, or any color other than blue, may be brought back to laying blue eggs in a few generations by persistently incubating only eggs fertilized by a cock that hatched from a blue-shelled egg, then incubating eggs from his offspring that have the bluest shells. Regardless of the shell color of the eggs your hens lay, if the color of your future hens' eggs is important to you, hatch eggs only of the desired hue or shade.

Can You Really Sex an Egg?

Wouldn't it be nice to save incubator space by hatching eggs that produce mostly pullets? Unfortunately, despite old wives' tales to the contrary — that cockerels hatch from elongated or pointy eggs or eggs over which a key swings longitudinally, while pullets hatch from round eggs or those over which a key swings crosswise — no sure way has been found to determine the sex of a fertilized egg or a developing embryo by examining the outside of the egg. (See my take on these superstitious practices in Screwpot Notions in the appendix.)

If determining in advance which gender will hatch from a given egg were a simple matter, the poultry industry wouldn't persist in spending time and money seeking a way to sex eggs prior to hatching them. To date, the best they've come up with is a tedious and expensive process involving inserting a needle into each incubating egg to remove and test a sample of fluid to determine whether or not it contains the estrogen compounds found within developing female eggs.

A RAINBOW OF SHELL COLORS

Hatching a Double Yolker Is a Rarity

Overly large eggs often contain two yolks, and double yolkers rarely hatch successfully because the shell is not roomy enough to accommodate two embryos. Assuming two do begin to develop, with insufficient space for growth, one or both will likely be deformed; one is apt to be considerably smaller than the other; and both, or at least the smaller of the two, may run out of growth space or oxygen and die in the shell. If both make it to hatching time, the cramped quarters would hamper the gymnastics required for them both to break into the airspace and eventually escape from the shell.

Despite all these obstacles, on rare occasions a pair of twins manages to hatch from a double-yolk egg, although typically one or both will be physically challenged and may not survive.

Egg Sources

Many of the same sources that offer chicks, described under Finding the Hatchlings for You on page 23, also sell hatching eggs. Some of the most popular sources are mail-order hatcheries, fellow poultry keepers, and specialty breeders. Some sellers have their own websites, while others sell on eBay or on any of the several poultry auction sites. Sources for hatching eggs abound on the Internet and may be found by doing a keyword search for "hatching eggs."

The one place you don't want to acquire eggs for hatching from is the grocery store. Eggs sold there are not fertile, but even if they were, they wouldn't hatch because they typically have been stored too long to remain viable, and at any rate they have been refrigerated. If you have access to a farmers' market or other place where fertile table eggs are sold, you may be able to talk the seller into saving some nonrefrigerated eggs for you to hatch. However, such sources are likely to have highly productive sex-link hybrids that will not produce chicks that are anything like the hens that laid the eggs.

No Guarantees

Regardless of where you get your hatching eggs, except maybe from a friend, relative, or neighbor, they are unlikely to come with any kind of guarantee. Even if the eggs are perfectly viable at the time you obtain them, incubation is tricky business — whether in a mechanical incubator or under a hen — and any number of things can go wrong between day one and hatching time. Once the eggs are out of the seller's hands, that person has no control over what happens to them. So purchasing eggs for hatching is always a gamble, and just as with any form of gambling, it's best not to put more money into it than you can afford to lose.

SHIPPED EGGS: A RISKY INVESTMENT

When you buy eggs from someone you don't know or who is not known to the greater poultry community, you never know what you're getting. The breeder flock may be unhealthy or of poor quality. The eggs may have been stored improperly, stored for too long, or packed insufficiently to protect them in their journey through the postal system. The eggs may not even be fertile.

So the first thing to do before buying online is to check the person's reputation by looking at auction-site feedback ratings and checking poultry forums, which occasionally discuss sellers to be avoided. Once you have identified reputable sellers, check to see how close they are to you. The less time eggs spend in transit, the better chance they have of producing chicks. So look for a seller as close to you as possible. You might get lucky and find someone near enough for you to pick up the eggs in person. Not only will that reduce the eggs' travel time, but also you will avoid the sometimes-exorbitant cost of having them shipped.

Eggs that are shipped can encounter any number of problems that will affect their hatchability. Here are some of the more common issues:

- Poor packing, which can result in cracked or broken eggs
- Rough handling, including bouncing around in a rural carrier's delivery vehicle
- Being subjected to summer heat or winter freezing
- Being transported in a plane's nonpressurized hold
- Remaining in transit for too long

If at all possible, eggs should be shipped on a Monday for delivery in midweek. Eggs shipped later in the week run the risk of a weekend layover, perhaps under conditions that are less than optimal to maintain good hatchability.

For a discussion on the effect of security scans on the hatchability of shipped eggs, see Screwpot Notions in this book's appendix.

When the eggs arrive, allow them time to settle before placing in your incubator. Settling has two purposes: to give the eggs a rest after being jostled during transport and to give them time to adjust to a uniform temperature before incubation starts. Let the eggs settle for at least half a day in an environment that is ideal for hatching-egg storage, as described on page 155. You will likely not get the same hatch percentage as you would get from fresher eggs that weren't subjected to the rigors of transport, but settling should improve your percentage rate compared to eggs that haven't been settled.

Some years ago a friend from a different state and I both planned to visit the same farm show, so he brought me one dozen hatching eggs of a breed I had long been seeking. I was pretty excited about the possibility that those eggs might hatch, but I didn't get my hopes up because I knew the eggs would be making a long journey to get to the farm show, sit around a motel for several days during the show, then make another long journey to my place. On the way home I insulated the eggs to keep them from getting cooked by the summer heat beating down on my vehicle while I stopped for lunch, but I learned that my friend had carried them on the seat of his pickup without any protection from heat or direct sun. I gave the eggs time to settle before placing them in the incubator. Three of the 12 eggs produced chicks and — how lucky can you get — they turned out to be two pullets and a cockerel.

EGGS FROM YOUR OWN FLOCK

Collecting hatching eggs laid by your own hens is the best way to ensure they are fresh and handled properly to maintain viability, as well as to avoid introducing some disease. Collect eggs at least three times a day to minimize their contact with dirty surfaces and to keep them from getting chilled or overheated, which reduces their hatchability.

Among all species, as hens age the number of eggs laid per year gradually declines, egg size gradually increases, and hatchability tends to decrease gradually. Likewise, male fertility gradually declines with age.

However, hatching eggs laid by hens that are at least two years old gives you plenty of time to evaluate the flock's track record, and hens that are still laying well at two years are likely to pass along to the next generation not only their laying ability but also their vigor and longevity. Although egg production declines with age, hens that lay well during their first and second year will continue to pass their superior qualities along to their chicks.

If you are intent on producing future generations of poultry, and especially if you are involved in breed preservation, you need a working knowledge of breeding methods, including linebreeding, pen breeding, double mating, pedigreeing, and progeny testing. Also helpful is a rudimentary knowledge of poultry genetics, lethal genes, and inbreeding depression. These topics, along with details on breeder flock feeding and management, are discussed in detail in *Storey's Guide to Raising Chickens*, 3rd Edition (2010).

Don't Save the First Egg

As exciting as finding a young hen's first egg can be, it is less likely to hatch than subsequent eggs. The same is true of any egg laid at the start of a new clutch, regardless of the hen's age.

Various reasons have been suggested, including that the first egg in each clutch remains inside the hen for a longer period between maturity of the yolk and being laid, during which the hen's body warmth causes premature embryo development. Another suggestion is that first eggs tend to be infertile. The most intriguing possibility relates to asexual embryo development, as described on page 185.

Clean Eggs Are a Must!

Avoid incubating dirty eggs. An incubator's operating temperature and humidity provide an ideal environment for the growth of bacteria, which can infect developing embryos or cause eggs literally to explode during incubation. Believe me, you don't want to experience the stench and cleanup job that follows the explosion of a rotten egg inside your incubator.

KEEP A SANITARY ENVIRONMENT

Manage your flock so eggs are clean when you collect them, which includes keeping floor litter clean and dry so hens won't track muck or mud into their nests. Discourage hens from laying on the floor by providing nests that are raised off the floor. When nests must be reached by means of a short ladder or at least a perching rail, soil is less likely to cling to feet.

Frequently remove any messy nesting material, and if debris remains clinging to the nest, scrape or scrub it out. If eggs consistently get dirty in the nest, reconsider your nest maintenance and the frequency with which nesting material is replaced. A lot of people, including me, prefer clean, dry wood shavings as nesting material, but straw or hay works just as well. A handy option is excelsior nest pads, manufactured from wood shavings and brown paper, and sold by some poultry suppliers. If you choose to use them, keep extra pads on hand to replace dirty ones. Introduce pads at the pullet stage, since hens that are accustomed to nesting in one kind of bedding may not take readily to something new.

Provide at least one nest for every 4 or 5 hens. The coop housing for my largest layer flock of some 40 hens has 10 nests. The coop that houses my half-dozen Silkie bantams has three nests.

Furnishing too few nests causes hens to crowd together into nests, or at least to accumulate too many eggs in a single nest, which can lead to broken eggs. Eggs that break in the nest contaminate unbroken eggs and render them unfit for hatching.

Too-thin shells is another reason eggs break in the nest. Thin shells may cover a pullet's first few eggs or the eggs of a hen that's getting on in age. A

Select eggs from your flock that are of a normal size, shape, and color for your species and breed.

pullet that isn't yet fully geared up for egg production lays eggs with thin shells. In an aging hen, the same amount of shell material (or less) that once covered a small egg must now cover the larger egg laid by the older hen, stretching the shell into a thinner layer.

Shells are generally thicker and stronger in winter and early spring but thinner in warm weather. Thin shells also may be due to a hereditary defect. If only one hen is laying thin-shell eggs, the best solution is to remove that hen from the breeder flock. If several or all hens in the flock consistently lay thin-shell eggs, the problem may be an imbalanced ration (too little calcium or too much phosphorus) or some disease.

An egg with a thin shell that breaks in the nest is usually eaten, either by the hen that accidentally broke it, or the next hen that comes along to lay in that nest. When a hen finds out how tasty eggs are, she'll start pecking them open on purpose to eat them and may teach other hens to do the same. The best solution is to remove the culprit early. If you can't catch the instigator in the act, you can often identify her by checking beaks for smeared egg yolk.

Shell-Building Supplements

Calcium is needed by laying hens to keep eggshells strong. The amount of calcium a hen needs varies with her age, diet, and state of health; older hens, for instance, need more calcium than younger hens. Hens on pasture obtain some amount of calcium naturally, but illness may cause a calcium imbalance. In warm weather, when all chickens eat less, the calcium in a hen's ration may not be enough to meet her needs, and a hen that gets too little calcium lays thin-shelled eggs. On the other hand, a hen that eats extra ration in an attempt to replenish calcium gets fat and becomes a poor layer.

Eggshells consist primarily of calcium carbonate, the same material found in oyster shells, aragonite, and limestone. All laying hens should have access to a separate hopper full of crushed oyster shells, ground aragonite, or chipped limestone (not dolomitic limestone, which can be detrimental to egg production).

Phosphorus and calcium are interrelated — a hen's body needs one to metabolize the other. Range-fed hens obtain some phosphorus and calcium by eating beetles and other hard-shelled bugs, but they may not get enough. To balance the calcium supplement, offer phosphorus in the form of defluorinated rock phosphate or charcoal (biochar). The correct ratio of phosphorus to calcium is 1:2. When both supplements are offered separately and are available at all times, hens will ingest the right balance.

CLEANING DIRTY EGGS

Slightly soiled eggs destined for incubation may be dry-cleaned with fine sandpaper or a dry sanding sponge. You don't want to wash the eggs, for fear of removing the protective bloom, but you might dip them into a sanitizer. Some poultry keepers routinely sanitize all hatching eggs, which does not affect the bloom and does improve the hatch rate. Sanitizing should be done as soon as possible after eggs have been collected from the nest.

A handy sanitizer may be made by adding 1 teaspoon of household chlorine bleach (Clorox) to 1 quart (1 L) of water warmed to 101°F (38°C). Never use water that's cooler than the eggs, because it can force contamination through the shell. Dip the eggs in the sanitizer for at least 1 minute but no more than 3 minutes, then set them on a clean towel to air dry; do not rub them dry, which would rub off the bloom.

Many products are available for sanitizing eggs, including quaternary ammonia products (such as Germ-X), phenol based products (such as Tektrol), and 1 percent iodine solutions (such as Durvet). These sanitizers vary in their friendliness to the chicken keeper, developing embryos, and the environment.

Rather than use chemicals I prefer to keep nests clean and discard dirty eggs, and I usually get respectable hatching rates (sometimes nearly 100% of fertiles). These days I hatch eggs only from my own hens. If I were still hatching eggs from outside sources, as I once did, I would sanitize them to avoid exposing my chicks to disease.

Egg Candling

Not all of an egg's qualities may be determined by looking at the outside of the egg. Detecting such things as hairline cracks, thin spots, and double yolks requires candling the eggs. Candling means examining the contents of an egg by placing a bright light behind it, although I'd be surprised if anyone still uses candles.

CANDLING DEVICES

Poultry-supply outlets offer various handheld devices designed specifically for the purpose. Most of them look like small flashlights with a plug-in cord. Although battery-operated candlers are available for candling eggs under a broody hen when no power is available, a small flashlight works at least as well as, if not better than, either a plug-in or battery-operated candler.

All you need is bright light that comes through an opening smaller than the diameter of the eggs you want to candle. If you have a too-big flashlight, cut a hole in a piece of cardboard and tape it over the business end of the flashlight, or tape a short piece of empty toilet-paper or paper-towel tube to the end in such a way that light only comes through the narrowed opening.

Such candling devices work best in a dark room. Hold the egg at a slight angle, large end to the light. Making sure your fingers don't block the light, turn the egg until either you see something or you're certain there's nothing to see.

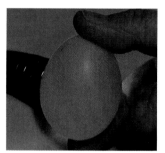

Candling may be done with a small, powerful flashlight in a darkened room.

More expensive, but fun and easy to use, is a battery-operated OvaScope consisting of two parts: a tabletop egg candler fitted with a black plastic hood that looks something like a microscope. Although the candler may be used alone, with the scopelike hood surrounding the egg and candler, ambient light is blocked to make the contents easier to view — which means that with the OvaScope you don't have to candle your eggs in a closet. The candling cylinder on which the egg is mounted has an adjustment wheel that lets you rotate the egg while you're examining it. The scope also slightly magnifies an egg, which is handy when candling smaller eggs.

EYE PIECE

WEBCAM ATTACHMENT

EGG ROTATION WHEEL

POWER SWITCH

An egg candled in the OvaScope (shown here with the front panel removed) may be viewed either through the eye piece or through a webcam attached to the eye piece.

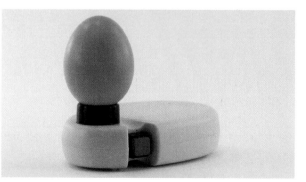

The base of the OvaScope (without the hood) may be used as a table-top egg candler in a darkened room.

WHAT TO LOOK FOR WHEN CANDLING

White-shelled eggs are easier to candle than eggs with colored shells, which is why white eggs have become the industry standard. Similarly, plain-shelled eggs are easier to candle than eggs with speckled shells. Double yolks are more difficult to see in colored or speckled eggs, but the unusually large size of a double yolker is already a pretty good indication of what's inside.

A blood spot is also difficult to detect in an egg with a speckled shell. It appears as a small, dark dot on the egg yolk, or sometimes within the white. A blood spot is not harmful, but is certainly not appetizing to find in a boiled or fried egg. Since blood spots can be hereditary, watch for them if you're hatching replacement layers for table egg production.

Hairline cracks and other shell imperfections are easy to spot on any egg. Cracks appear as white veins in the shell. Since cracks open the way

This shell has a clearly visible crack, and candling reveals why it cracked: the shell is extremely thin and patchy.

for bacteria to enter, eliminate these eggs from your selection for hatching. Cracks are more likely to occur in eggs with thin shells.

AGING EGGS

Besides letting you easily see shell imperfections and double yolks, candling lets you estimate an egg's age, which is important if you discover a hidden nest full of eggs and have no idea how long they've been there. The size of an egg's air cell increases as an egg ages. A freshly laid egg has no air cell at all, but as the egg cools and its contents shrink, the air space develops. Then, as moisture evaporates from within the egg, the inner shell membrane pulls away from the outer shell membrane, and the air cell grows.

Just how fast the air space grows depends on the porosity of the shell and on the temperature and humidity the egg has been subjected to. The cell of a freshly laid cool egg is no more than ⅛ inch (3 mm) deep. From then on, the larger the cell, the older the egg. In a nest full of found eggs, you will see a progression in air cell sizes based on how long ago or how recently each egg was laid.

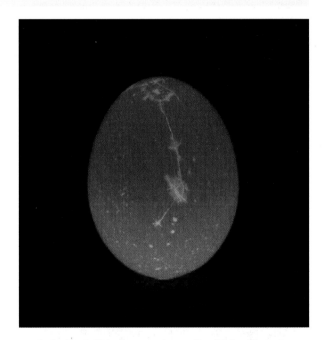

Cracks in the shell that may not be readily visible will appear during candling as white veins and/or spiderweb-like shapes.

Egg Storage

An egg need not be rushed into the incubator the moment it is laid. Saving up eggs is, in fact, a natural part of the incubation process. In nature, eggs remain dormant while a hen accumulates a full setting before she starts to brood. That way the embryos can all develop at the same rate and hatch at the same time, ensuring that late hatchers aren't left behind when early hatchers are ready to leave the nest.

But even under optimum storage conditions, the longer eggs are stored, the longer they will take to hatch. Also their ability to hatch decreases with time. When eggs are stored prior to incubation, you can expect the following effects:

- A 40-minute increase in incubation time per day of storage
- A 0.5 to 1 percent decrease in hatchability per day of storage

You can store eggs for up to 6 days without noticing a significant difference. I don't feel

Temperature Shock

Hatchability can be affected by temperature shock, which is caused by a too-rapid change in the temperature of hatching eggs. It most commonly occurs when eggs are sanitized at temperatures that are too hot or too cold, when eggs are exposed to temperature variations during transport, when eggs are put into or taken out of storage, or when cool eggs are placed in a warm incubator. Eggs brought out of cool storage should be allowed to warm to room temperature before being placed in the incubator.

comfortable storing them for longer than 10 days, although on occasion I've had decent hatches from eggs stored as long as 14 days when it took that long to collect enough to be worth running the incubator.

STORAGE CARTONS

To avoid contamination during storage, place the eggs in clean cartons. Hard-plastic egg cartons, of the sort backpackers use, may be easily disinfected for reuse. Recycled Styrofoam or cardboard cartons from the grocery store accumulate bacteria over time. When I use recycled cartons to store hatching eggs, I discard them after one use. (For my take on sanitizing egg cartons in the microwave, see Screwpot Notions in this book's appendix.)

Place eggs in the carton with their large ends up to keep the yolks centered within the albumen. If the eggs will be stored longer than 6 days, keep yolks from sticking to the inside of the shell by tilting the eggs from one side to the other. Instead of handling eggs individually, elevate one end of the carton one day, and the opposite end the next day.

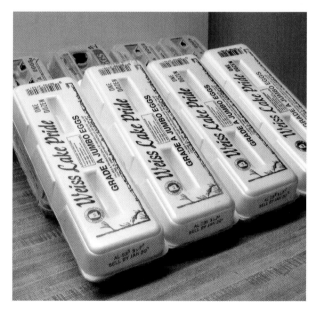

Eggs held longer than 6 days before incubation should be stored with one end of the cartons elevated one day, and the opposite end elevated the next day.

STORAGE CONDITIONS

When a hen selects a place to make her nest, she instinctively seeks out conditions that are optimum for them to remain viable until she is ready to start setting. When you store eggs collected for hatching, look for a place in your house that duplicates those conditions as closely as possible. It might be a dry cellar, a cool back room closet, or a food storage pantry. In a cool climate the concrete floor of a garage or basement may be the ideal place to store hatching eggs, provided the eggs are protected from being mauled by pets or eaten by predators.

Keep the eggs out of sunlight in a cool, relatively dry place but not in the refrigerator. The ideal temperature is 55°F (13°C), but anywhere between 50 and 60°F (10 and 15°C) is adequate.

The ideal humidity is about 75 percent, with a range of 70 to 80 percent. At a much higher humidity, moisture condensing on the shells would attract harmful molds. High humidity also encourages any bacteria on the shell to multiply.

Excessive dryness, on the other hand, increases the rate at which moisture evaporates through the shells. The less moisture that evaporates from eggs during storage, the better chance they have of hatching. Small eggs laid by bantams and jungle fowl have a relatively large surface-to-volume ratio and therefore evaporate more quickly than larger eggs. Late-summer eggs of any size have thinner shells because the hen has been calling on her calcium reserves all summer. These shells allow more rapid evaporation than that which occurs in early-season eggs.

To minimize evaporation of eggs stored longer than 6 days, seal cartons in plastic bags, or better yet, wrap the eggs individually in plastic wrap. Carefully wrapped eggs stored under optimal conditions will maintain reasonable hatchability for as long as 3 weeks.

9

OPERATING AN INCUBATOR

When you're ready to start hatching, set up your incubator and run it for at least half a day so the temperature and humidity have time to stabilize before you put in any eggs. For the steadiest incubation temperature, locate your incubator where the room temperature remains fairly constant. The ideal room temperature is between 75 and 80°F (24–27°C), although anywhere between 55 and 90°F (13–32°C) is acceptable. Position the incubator away from windows, where temperature may be affected by sunlight falling on the incubator during the day or cold air seeping through the window at night. Similarly, keep the incubator away from heating and air-conditioning vents that periodically blast hot or cold air.

Adjusting Temperature

Operating an incubator at the correct temperature can be tricky. Unless the incubator is extremely well insulated, temperature fluctuations in the room where the incubator is set up can affect the hatch. A thermometer that isn't properly positioned according to the incubator manufacturer's instructions can give a false temperature reading.

Even if the thermometer is properly placed, incubation temperature can't be accurately monitored if the numbers on the thermometer are too close together, as is often the case with a dial or linear thermometer. For this reason a digital model is a good choice. The thermometer itself may be inaccurate; it's always wise to check a new thermometer against a second or even a third thermometer that you're sure is accurate.

A typical operating temperature for a forced-air incubator is 99.5°F (37.5°C); for a still-air incubator, 102°F (39°C). The actual amount of heat reaching the developing embryos inside the eggs is the same for all incubators. The differences in temperature, as read on the thermometers of different incubator models, have to do with their specific thermostat, the location of the eggs and proximity of the thermometer to the eggs, and differences in their systems for ventilation and heat distribution. All these factors combine to ensure that, when incubators are operated according to their manufacturers' instructions, the temperature reaching the embryos is the same in all incubators.

Talking Fowl

dry-bulb temperature. The temperature of air as measured by a thermometer that is freely exposed to the air; the dry-bulb temperature in a properly maintained incubator is higher than the wet-bulb temperature

help-out. A bird that is unable to hatch from an egg without assistance

hygrometer. An instrument for measuring the amount of moisture in the air in either wet-bulb degrees (as measured by a wet-bulb thermometer) or relative humidity percentages (as measured by a digital hygrometer)

pip. The hole a newly formed chick makes in its shell when it is ready to hatch. Also the act of making such a hole

set. To place a group of eggs together in an incubator or under a hen

setting. A group of eggs placed together in an incubator or under a hen

water pan. A shallow container in an incubator, from which water evaporates to provide humidity; also called evaporation pan

wet-bulb temperature. A measure of the amount of moisture in the air, as determined by fitting the end of a thermometer with a moistened wick from which water evaporates; the wet-bulb temperature in a properly maintained incubator is lower than the dry-bulb temperature

Make It Easy to Remember

Put the incubator in a place that is convenient for you to check often, but not in a high-traffic area. Next to a doorway where people and pets are constantly coming and going is not a good idea if the plug might accidentally be pulled from the wall or young children might be tempted to play with the thermostat. On the other hand, a rarely used back bedroom where you'll forget to check your incubator might not be so great, either. You'll want your incubator in a convenient place where you'll remember to monitor temperature, humidity, and ventilation and to turn the eggs — all of which affect the hatch.

If the incubator is run at a temperature that's 0.5 to 1°F (0.3–0.5°C) too high or too low, the eggs will not hatch at the appropriate time. Eggs that pip may fail to hatch, and in general the hatch rate will be lower than if the temperature had been just right.

When hatchlings emerge late, incubation temperature is low. The hatchlings tend to be big and soft with unhealed navels, crooked toes, and thin legs. They may grow slowly or may not eat and drink at all.

When hatchlings emerge early, incubation temperature is high. The hatchlings tend to have splayed legs and can't walk properly.

Lethal Temperatures

Several hours' incubation at too high a temperature is much more damaging than the same amount of time at a temperature that's too low. In a still-air incubator, embryos will die at 107°F (41.7°C). The lethal temperature in a forced-air incubator is 103°F (39.5°C).

When some eggs hatch on schedule, but not all pips hatch, and the hatch rate of fertile eggs is poor, the problem is not the temperature. Rather, most likely the humidity is off.

TEMPERATURE AND HUMIDITY

Optimum incubation temperature and humidity are interrelated. As the temperature goes up, relative humidity must go down to maintain the same hatching rate. Here are some typical combinations for a still-air and a forced-air incubator:

Still-Air
102°F (38.9°C) at 58 percent humidity
100°F (37.8°C) at 61 percent humidity

Forced-Air
99°F (37.2°C) at 56 percent humidity
98°F (36.7°C) at 70 percent humidity

When you operate any incubator for the first time, follow the manufacturer's recommendations as closely as possible. For future hatches you'll most likely have to make minor adjustments to improve your success rate. Keep accurate records (as outlined on page 181, next chapter), and after a few hatches you will hit on the optimum settings for your particular circumstances. For example, small eggs evaporate more rapidly than large eggs, so they hatch better at a lower temperature and higher humidity.

POWER FAILURE? POSSIBLE FIXES

The more valuable your hatching eggs, the more you can count on the power going out at a critical time during incubation. I live in a rural area and frequently experience power outages during spring thunderstorms at the height of our hatching season. After losing too many hatches to extended power outages, we got a portable standby generator.

If you don't have a generator and the power stays off for any length of time, open the incubator and let the eggs cool until the power goes back on. Cooling reduces embryo metabolism and slows growth, while trying to keep the eggs warm can cause abnormal development. Furthermore, if you close the vents or wrap the incubator with blankets, you run the risk of oxygen deprivation, a greater danger than heat loss. In a prolonged outage the oxygen level could fall below that necessary to keep the embryos alive.

As soon as the power goes back on, close the incubator and continue normal operation. The effect of the outage on your hatch will depend on how long the eggs had been incubated before the outage and how long the power was out. A power failure of 18 hours will delay the hatch by a few days and significantly reduce the success rate.

An outage of up to 12 hours may not significantly affect the hatch — except to delay it somewhat — especially if the outage occurred during early incubation, when cooled embryos naturally go dormant. Embryos that are close to hatching may generate just enough heat to carry them through a short-term outage. (For my take on using an inverter to plug your incubator into your vehicle's cigarette lighter, see Screwpot Notions in the book's appendix.)

If your incubator is electronic, you should protect it with a surge suppressor. Frequent power spikes (brief increases in voltage) or surges (longer voltage increases) are all too common and can wear down an incubator's electronic system. A high-voltage spike or surge can end your hatching season in a hurry.

Where electricity is iffy or not available, a nonelectrical option is an Amish-made kerosene incubator. Operating one can be tricky business for someone who is not thoroughly familiar with using oil lamps and wicks, but a kerosene incubator is the ideal solution for anyone living off the grid.

Adjusting Humidity

Humidity is a measure of the amount of moisture in the air. At a given temperature, the drier the air, the more rapidly the air absorbs moisture. The more moisture the air already holds, the less additional moisture it can absorb. Applying this concept to incubation, the drier the air in the incubator, the more rapidly moisture will evaporate from the eggs through their shells. The moister the incubator's air, the more slowly moisture will evaporate from the eggs.

For a successful hatch, moisture must evaporate from eggs at just the right rate. Incorrect humidity can cause embryos to die in the shell or result in abnormal hatchlings.

When incubation humidity is too low, embryos in the early stage of incubation may adhere to the shell membrane and die. Those that continue to grow may be small and too weak to pip, may pip and not hatch, or may crack the shell all the way around but be unable to free themselves from a dried-out membrane. If a hatchling manages to struggle free, or you help it out, it may have a twisted neck or crooked feet, or be unable to stand.

Low humidity tends to be a problem with a small incubator that has an inadequate water container, a large incubator that is not filled to capacity, and any incubator that must be frequently opened to turn the eggs manually. A sudden drop in humidity during a hatch may mean the water pan is coated with chick fluff, preventing evaporation and causing the humidity to plummet.

When incubation humidity is too high, embryos grow too large inside the shell to move into hatching position, and if they do manage to break into the air cell, it is too small to provide sufficient oxygen, and the term embryo will suffocate. Birds may not hatch because the shell membrane is too rubbery to punch through. Those that do hatch will be large, soft, and sluggish.

Excessive incubation humidity is usually due to insufficient ventilation or high ambient humidity. An accumulation of moisture on the incubator's observation window during the hatch is an obvious indication of excess humidity.

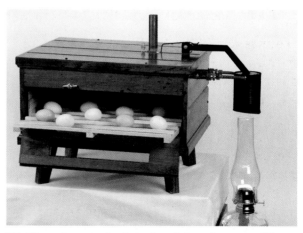

This 50-egg kerosene incubator from Lehman's uses an oil lamp to heat water that circulates in copper tubing to maintain a constant hatching temperature.

MORE HUMIDITY FOR THE HATCH

The requisite amount of humidity is not steady from start to finish but is higher during the hatch than during incubation. In nature the first chicks to hatch contribute to humidity under the hen, thus assisting the later, and typically weaker, hatchlings to get free of their shells.

An incubator that calls for about 60 percent relative humidity during incubation should be increased to 70 percent for the hatch. A wet-bulb reading that is 86 to 88°F (30–31°C) during incubation typically should be increased to 88 to 91°F (31–33°C) during the hatch.

Small eggs, such as those laid by bantams and guineas, as well as waterfowl eggs, generally need more humidity at hatch than other eggs. All eggs require more humidity as summer progresses and shells become more porous. Adequate humidity during the hatch prevents the shell membranes from drying out and sticking to the emerging embryos, a condition that effectively puts them into a straitjacket and prevents them from getting out of the shell.

HUMIDITY REGULATORS

An incubator's humidity is regulated by a number of different factors you can't control, and a few that you can.

Water pan. The level of humidity is largely regulated through evaporation from a container of water, called a water pan or evaporation pan. In a large incubator the water pan is usually removable; in a small incubator it may be molded into the incubator's bottom. The pan's surface area determines how rapidly water evaporates from the pan, thus regulating the amount of humidity in the incubator — the larger the surface area, the more moisture can evaporate; the smaller the surface area, the less moisture can evaporate.

Surface area is decreased by partially covering the water pan with foil, by removing one or more pans, or removing water from sections of a single pan. Surface area is increased by adding pans, by adding water to additional sections of a single compartmentalized pan, or by vertically inserting sponges (humidity pads or wick pads) into the pan.

Ventilation. Opening vents decreases humidity by allowing moist air to escape; closing vents increases humidity by trapping moist air in the incubator.

Capacity. The eggs themselves contribute to humidity by evaporating through the shell. Whether or not the incubator is filled to capacity therefore influences the humidity level.

Ambient humidity. Where the ambient humidity is constantly variable, it's a good idea to monitor the humidity level in the room where your incubator is located using an inexpensive thermometer that measures both temperature and humidity. Where the room is not air-conditioned, you might just check the local weather report. If the ambient humidity is 15 percent or more than the desirable incubation humidity, remove the water pan. Be sure to replace the water pan as soon as the ambient humidity falls below that level.

Those of us who live where the weather is extremely humid half the year and extremely dry the other half must compensate accordingly. At the beginning of the season, when spring rains result in a high ambient humidity, I have to be careful to avoid excessive humidity in the incubator. As fall approaches and my hatching season winds down, ambient humidity drops at the same time my cabinet incubator contains progressively fewer eggs after each hatch, so I have to take care to boost the humidity to ensure a successful final hatch.

Adding Water

Since cool water draws heat from the incubator, always use room temperature or lukewarm water when filling the water pan.

MEASURING HUMIDITY

Incubation humidity is measured either in wet-bulb degrees or as percent relative humidity. A wet-bulb reading is measured by a standard thermometer, the bulb of which is enclosed in a moistened wick so that water constantly evaporates from it and cools the bulb.

Relative humidity is the ratio of the partial pressure of water vapor to the saturated vapor pressure of water at a given temperature and is expressed as a percentage. Digital hygrometers measure percent relative humidity.

Take care not to confuse wet-bulb degrees with percent relative humidity. If your still-air incubator operating at 101°F (38.3°C) calls for 65 percent relative humidity, that would be 90°F (32.2°C) wet bulb. Trying to boost the humidity to 90 percent, or reduce it to 65°F (18.3°C) wet bulb, would be disastrous. The table below shows the equivalents between wet-bulb degrees and percent relative humidity at common incubation temperatures.

Factors Affecting Humidity

How rapidly eggs lose moisture, and therefore how much humidity is needed during incubation, is affected by a number of factors. They include:

- Egg size — the smaller the egg, the greater the percentage of moisture loss
- Shell porosity, which increases with hen's age
- Elevation
- Storage length and conditions
- Weather conditions
- Incubation temperature
- Air speed
- Shell thickness, which decreases with hen's age; thinner shells require higher humidity

Humidity at Common Incubation Temperatures

FORCED AIR TEMPERATURE				STILL AIR TEMPERATURE				RELATIVE HUMIDITY
Wet-Bulb Reading								
99.5°F/37.5°C		100°F/37.8°C		101°F/38.3°C		102°F/38.9°C		
80.8°F	27.1°C	81.3°F	27.4°C	82.2°F	27.9°C	83.0°F	28.3°C	45%
82.8	28.2	83.3	28.5	84.2	29.0	85.0	29.5	50%
84.7	29.3	85.3	29.6	86.2	30.1	87.0	30.5	55%
86.7	30.4	87.3	30.7	88.2	31.2	89.0	31.7	60%
88.5	31.4	89.0	31.7	90.0	32.2	91.0	32.8	65%
90.3	32.4	90.7	32.6	91.7	33.2	92.7	33.7	70%
91.9	33.3	92.5	33.6	93.6	34.2	94.5	34.7	75%
93.6	34.2	94.1	34.5	95.2	35.1	96.1	35.6	80%
95.2	35.1	95.7	35.4	96.8	36.0	97.7	36.5	85%

From: *Storey's Guide to Raising Chickens*, 3rd edition

Monitoring Air-Cell Size

In place of a hygrometer, or in conjunction with one, a good indication of humidity is the changing air-cell size inside the developing eggs, as determined by candling. Moisture evaporating from an egg causes its contents to shrink, which increases the size of the air cell. If air cells are proportionately larger than the one shown in the sketch opposite, increase humidity; if they're smaller, decrease humidity.

Since a hatchling breaks into the air cell before it breaks out of the shell, examining the shape of the shells after hatch can also give you clues about the humidity level. A hatchling should occupy approximately two-thirds of the shell and the air cell the remaining third. Humidity is correct when the bottom part of the shell is approximately twice as large as the cap. If the cap is much smaller than half, humidity is too high; if it's much larger than half, humidity is too low.

Where ambient humidity constantly varies, an inexpensive thermometer with a humidity indicator helps monitor room humidity.

Spraying and Cooling Waterfowl Eggs

A setting duck or goose leaves the nest to take a daily swim, during which her eggs are left to cool. When she returns, the residual moisture in her feathers dampens the eggs in the nest. This daily moistening of the tough shells encourages more rapid moisture loss from within the shells.

By imitating natural conditions as much as possible, you can improve your hatching rate. Briefly turn off and open the incubator daily from days 4 through 25, letting the eggs cool to about 86°F (30°C). If you don't have an infrared thermometer (temperature gun) to do the job, you can pretty accurately gauge an egg's temperature by touching the shell to your eyelid — if it feels neither warm nor cool, it's just the right temperature. Cooling takes 5 to 10 minutes. To make sure you don't forget and leave the eggs to cool for too long, which can be disastrous, plan to remain close by until cooling is complete or set an alarm to remind you.

Before closing the incubator and turning it back on, spray the eggs with lukewarm (100°F/°37.8°C) water. At the end of the turning period on day 25, or when the eggs are moved to a hatcher, spray them once more.

HOW HUMIDITY AFFECTS SHELL CAPS AND BOTTOMS

humidity too high

humidity just right

humidity too low

Humidity is correct when the bottom part of the shell after hatch is approximately twice the size of the cap

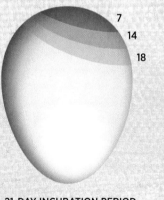

7
14
18

21-DAY INCUBATION PERIOD

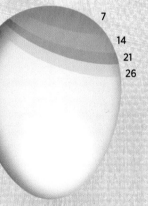

7
14
21
26

28-DAY INCUBATION PERIOD

At proper incubation humidity, the air cell size will increase at a steady rate and may be observed by candling.

Methods of Adjusting Humidity

TO INCREASE HUMIDITY ⬆	TO DECREASE HUMIDITY ⬇
Close adjustable vents	Open adjustable vents
Use a larger (or added) water pan	Use a smaller water pan
Add sponges (humidity/wick pads)	Partially cover water pan with foil

Monitoring Weight Loss

As moisture from within an egg evaporates during incubation, the egg's weight decreases. Under proper humidity an egg will lose 12 to 14 percent of its weight during incubation. So another way to measure humidity is to monitor weight loss as incubation progresses. At the start of incubation, weigh a sample of eggs — say, half a dozen — and weigh the same sample every 5 days. Then compare their actual weight to their expected weight based on average loss. If the actual weight is less than the expected weight, increase the incubator's humidity; if the actual weight is greater than expected, decrease the humidity.

The Expected Weight Loss during Incubation table below shows how much weight loss you can expect in a sample of 6 chicken eggs of differing sizes with typical starting weights for their size. Because of the small numbers involved, a scale that measures in the metric unit of grams is more accurate than a scale measuring ounces. Weight loss should be similar for eggs of other species except, of course, that the loss in weight would be stretched over the longer incubation period required for each species. A weight loss that is significantly greater or less than expected reduces hatchability.

One way to monitor humidity during incubation is to weigh a sample of eggs periodically.

Expected Weight Loss during Incubation of Chicken Eggs

DAY	0	5	10	15	21
Ounces per sample of 6 eggs					
Bantam	7.5	7.3	7.1–7.0	6.9–6.7	6.6–6.4
Small	9.0	8.7	8.5–8.4	8.2–8.1	7.9–7.7
Medium	10.5	10.2	9.9–9.8	9.6–9.4	9.2–9.0
Large	12.0	11.6	11.3–11.2	11.0–10.8	10.6–10.3
X-large	13.5	13.1	12.7–12.6	12.3–12.1	11.9–11.6
Jumbo	15.0	14.5	14.1–14.0	13.7–13.5	13.2–12.9
Grams per sample of 6 eggs					
Bantam	216	210–209	204–202	198–194	190–186
Small	258	251–249	243–241	236–232	227–222
Medium	300	291–290	283–280	274–270	264–258
Large	342	332–331	323–319	313–308	301–294
X-large	384	373–371	362–358	351–345	339–331
Jumbo	420	408–406	396–392	384–378	370–361

To determine if humidity level is correct, compare actual weight to the expected weight. If the actual weight is greater than expected, humidity is high; if the actual weight is less than expected, humidity is low. Expected weight for chicken eggs on any given day lies between $W - (0.0057 \times D \times W)$ and $W - (0.0067 \times D \times W)$, where W = starting weight and D = days of incubation.

From *Storey's Guide to Raising Chickens*, 3rd Edition

Ventilation Ensures Oxygen Supply

Ventilation involves replacing the air inside the incubator with air from outside the incubator. Ventilation is important because the surface of an egg's shell is covered with tiny pores through which oxygen enters and both carbon dioxide and moisture exit. Good incubator ventilation ensures a ready supply of oxygen for developing embryos (in other words, lets them breathe) and removes the carbon dioxide they produce. It also influences the degree of water evaporation from within the egg and, to some extent, removes excess heat produced by developing embryos.

AIR VENTS

A forced-air incubator uses a fan to stir up the air, which — as well as keeping the temperature uniform throughout the incubator — draws in fresh, oxygen-laden air through open vents. A still-air incubator works on the principle that cool air sinks, displacing warm air upward; as some of the warm air exits through vents at the top, cooler fresh air is drawn in through other vents.

In addition to vents that are open all the time to allow air exchange throughout incubation, some incubators have adjustable vents with either plugs that may be removed or covers that slide across the vents to let you fully or partially open them as incubation progresses. How many vents are closed or opened controls the amount of ventilation within the incubator.

Underventilating can result in too little oxygen and too much humidity. Overventilating can result in too little humidity, as well as difficulty in controlling temperature. Big industrial incubators automatically control ventilation by constantly sampling the air inside the incubator and making any necessary adjustments. We backyarders are pretty much left to wing it.

VENTILATION REQUIREMENTS

As the environment both inside and outside an incubator changes, so too does the need for more or less ventilation. Some of the factors affecting ventilation requirements include those in the following paragraphs.

Incubator size. The bigger the incubator, the more important ventilation control becomes. Since the need for ventilation control varies for different incubators, start out carefully following the manufacturer's recommendations and make future adjustments according to past hatching success and the variability of the other (following) factors. With a small incubator, controlling ventilation is generally not needed until hatching time — especially if the incubator is opened several times a day for egg turning.

Number of eggs set. Incubators are designed to operate at capacity. If fewer eggs are set than the incubator is designed for, the need for oxygen, production of carbon dioxide, amount of egg evaporation, and degree of heat production by developing embryos won't be as great.

Stage of incubation. At the beginning of incubation, developing embryos use little oxygen and produce little carbon dioxide. As incubation progresses and the embryos use increasingly more oxygen, they produce more carbon dioxide and also more body heat. The need for oxygen rapidly increases about two-thirds of the way into the incubation period and nearly doubles just before the hatch, making good ventilation essential at hatching time. Underventilation can cause embryos or hatchlings to die. On the other hand, overventilation can reduce humidity to the point that the hatch is delayed or reduced. Avoid this scenario by waiting until about half the eggs have hatched before significantly increasing ventilation. Adjusting ventilation as the hatch progresses is decidedly problematic in an incubator used for continuous hatching, which is one of the advantages of single-stage hatching.

Air movement in the incubator. In a small still-air incubator, especially one with an automatic egg turner (so that the incubator is not frequently opened for egg turning), more than the usual amount of ventilation may be needed to reduce humidity.

Conditions in the room. Ambient temperature and humidity can adversely affect incubator temperature and humidity. Where the room temperature is below 64°F (18°C), ventilation can reduce both temperature and humidity. Where ambient conditions are extremely hot and humid, the incubator may need to be situated in an air-conditioned environment for ventilation to be effective.

Altitude. The low atmospheric pressure at altitudes in excess of 3,000 feet (915 m) can result in insufficient oxygen and excess moisture loss from

Effect of High Altitude on Hatchability

ALTITUDE	HATCHABILITY
Sea level to 3,000 feet	Normal
3,000 to 3,900 feet	–5%
4,000 to 4,900 feet	–7%
5,000 to 5,900 feet	–9%
6,000 to 6,900 feet	–10%
7,000 to 7,900 feet	–12%

Adapted from "Factors Affecting Hatchability of Chicken Eggs," Bill Cawley, PhD, Texas A&M University

incubated eggs. The higher the altitude, the lower the hatchability. The problem becomes worse with eggs laid by hens acclimated to a significantly lower altitude. Decreasing ventilation reduces the problem of dehydration, but supplemental oxygen may also be needed to obtain a decent hatch rate.

Egg Position and Turning

How eggs are positioned in the incubator affects their hatchability. During all times the blunt end of the egg should remain higher than the pointed end. In an incubator with a turning device, the eggs are placed vertically in holding trays. In an incubator without a turning device, the eggs are placed on a horizontal tray, with care taken to keep the pointed end slightly lower than the blunt end.

Every now and then someone brings up the question of whether turning is really necessary. In an attempt to resolve the issue I incubated two identical settings under identical conditions, one turned as usual and the other not turned at all. The outcome in percentage of eggs that hatched

in each group was pretty dramatic in favor of turning. For more information on this subject see Screwpot Notions in this book's appendix.

During incubation the eggs must be turned mechanically or manually. The most critical time eggs need to be turned is during the first week of incubation. Signs of improper turning are early embryo deaths and full-term chicks that fail to pip.

The turning of eggs during incubation serves these important purposes:

- It prevents the embryo from sticking to the shell membrane.
- It helps even out the temperature within the egg.
- It moves metabolic wastes away from the embryo.
- By moving away wastes, it improves contact of the embryonic membrane with the albumen to improve access to the albumen's nutrients.

Wash and Wash Again

Always wash your hands thoroughly before turning eggs, and again afterward, to avoid transferring bacteria to or from the shell.

MANUAL EGG TURNING

In an incubator with automatic turning, the eggs are tilted or rolled from one side to the other, usually every hour. In an incubator without a turner, manual turning every hour isn't practical, but turning the eggs three or five times a day is nearly as effective. Since the most critical time for eggs to be turned is during the first week of incubation, you might turn them five times a day during the first week and then scale back to three times a day thereafter.

Unless you are available to turn the eggs at regularly spaced intervals around the clock, turn them an odd number of times each day so they are not always lying on the same side overnight. When I use a manual-turn incubator, I turn the eggs when I get up in the morning, approximately halfway through the day, and just before going to bed.

As with automatic turning, the eggs aren't turned end to end but from side to side. It's a good idea to leave enough space in the incubator so you can roll the eggs half a turn in one direction, then at next turning roll them half a turn back. Otherwise you'll have to lift and turn each egg individually, which is more time consuming and runs the risk of accidentally dropping an egg and breaking it (as well as the egg it lands on). Throughout incubation, the blunt end of the egg should always be oriented upward for a more successful hatch and stronger hatchlings.

Turn eggs only halfway around, not in a complete circle. Otherwise, you run the risk of breaking the allantois sac, which kills the embryo. To make sure each egg is turned halfway around, and all eggs are turned each time, place an X on one side and an O on the opposite side with a grease pencil or china marker. Some folks prefer a soft (number 2) pencil, but I stopped using a pencil after I pierced a few shells with the sharp

For manual turning, an X on one side of each egg and an O on the other side helps ensure that all eggs have been turned each time.

point. I prefer not to use a felt or ink pen because I worry about ink being absorbed through the porous shell.

Opening the incubator several times a day to turn eggs reduces the relative humidity within the incubator, which must be compensated for. One method is to spritz eggs with lukewarm water from a spray bottle. Depending on the design of your incubator, you might have to turn off the fan so it won't dry out the eggs by blowing air across them while they are being turned.

WHEN TO STOP TURNING

As hatching time approaches, turning should stop, to give the embryos time to get oriented and begin breaking out of their shells. Failure to stop turning the eggs can result in a poor hatch. Most incubation instructions say to stop turning 3 days before the hatch. Under natural conditions, the about-to-hatch chicks start peeping about 3 days before they hatch, causing the broody hen to stop fidgeting in the nest and thus stop turning her eggs.

If you use a hatcher or an incubator with a separate hatching tray, move eggs to the hatcher or hatching tray 3 days before they are expected to hatch. Be sure to continue turning any eggs remaining in the incubator or on the turning trays that are scheduled to hatch at a later date.

If you incubate and hatch in the same unit, you can stop turning chicken eggs after 2 weeks and the eggs of other species after 3 weeks. In a small

When to Stop Turning

TYPE OF EGG	STOP TURNING ON DAY
Chicken	19
Duck, Guinea, Goose, Turkey	25
Muscovy	31
Bantam, small	17

incubator with a turner that has to be removed before the hatch, you can remove the turner any time between 2 weeks (for chickens, 3 weeks for others) and 3 days before the hatch. In an incubator with a turning rack, disconnecting the turner but leaving the eggs to hatch on the turning trays is not a great idea, as hatchlings can be pretty active, and you don't want them to get tangled up in the rack and possibly twist or break a leg.

Embryo Development

During incubation embryos go through many stages, as measured by the development of each body part and system. These stages occur much more rapidly during early incubation. About two-thirds of the way into the incubation period, the embryo is basically fully formed, and development continues primarily as an increase in size. By the day of hatch, the embryo completely fills the shell except for the air cell.

RATE OF DEVELOPMENT

Not all eggs go through the stages at precisely the same rate. Of course, embryos of different species develop at differing rates. But even among eggs of the same species, embryonic development may vary. Some of the factors that influence the rate of development include these:

- Genetic differences — some breeds, or strains within breeds, develop more or less rapidly than the norm for the species
- Stage of development prior to incubation — some development may occur under a hen spending time in the nest

- Length of storage prior to incubation — the longer eggs are stored, the slower the embryos develop
- Temperature of eggs when placed in incubator — cool eggs take longer to warm up, delaying development
- Egg size — smaller eggs than the norm hatch faster, larger eggs take longer
- Season — eggs generally have greater nutritional value in spring, causing embryos to grow more vigorously
- Temperature and other factors related to the environment within the incubator

Embryonic development actually starts while an egg is being formed inside the warm body of a hen, during which a fertilized egg begins the process of cell division. As the egg cools, embryonic development becomes suspended, which is why fertile eggs may be held for a few days at about 50°F (10°C) prior to being incubated. However, temperatures of 80°F (27°C) or greater for an extended period cause slow cell division until eventually the embryo becomes weakened and dies.

SUPPORT SYSTEMS

Starting on the first day of incubation, four systems of membranes develop inside an egg that help the embryo develop, live, and grow. Because they do not become part of the hatchling's body, they are called extraembryonic membranes. These support membranes are as follows:

Yolk sac or vitelline sac. This membrane surrounds the yolk and is where the first blood vessels form during the first day of incubation. Its function is to transport nutrients from the yolk to the embryo. As the embryo grows and the yolk gets used up, the membrane gets smaller.

Allantois. This sac arises from the embryo's gut, starting on the third day of incubation. The allantois is similar to the yolk sac, and in fact the two systems work together to function much like a mammal's umbilical cord. As the chick embryo develops, using up nutrients from the yolk and expelling wastes into the allantois, the allantois grows until it completely surrounds the embryo. Its function is to provide the embryo with oxygen, expel carbon dioxide, deliver nutrients from the albumen and calcium from the shell, and collect body wastes (the greenish gunk found on the hatching tray at the end of incubation).

Amnion. This innermost membrane begins developing on the third day of incubation and surrounds only the embryo. It is filled with colorless amniotic fluid in which the embryo floats — a stable environment that gives it freedom to move and exercise, protects it from jarring impacts, and prevents it from drying out.

Chorion. This membrane is an overall enclosure that surrounds and protects the embryo and its support systems. As the allantois grows within the chorion, it eventually touches the chorion near the egg's blunt end and the allantoic membrane begins to fuse with the chorion. By the time the allantois surrounds the embryo, the allantoic membrane and the chorion membrane have become one and the same. This chorio-allantoic membrane keeps growing, until by the sixteenth day of incubation it nearly fills the inside of the shell. It stops functioning just before the hatch, as soon as the embryo breaks into the air cell and begins using its lungs to breathe.

EMBRYONIC SUPPORT MEMBRANES

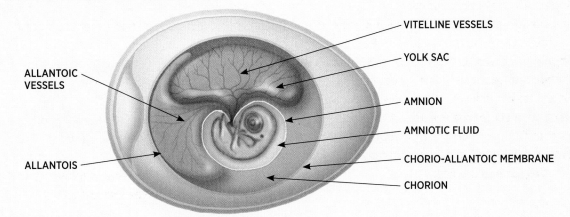

ALLANTOIC VESSELS

ALLANTOIS

VITELLINE VESSELS

YOLK SAC

AMNION

AMNIOTIC FLUID

CHORIO-ALLANTOIC MEMBRANE

CHORION

Stages of Incubation

Although development occurs in specific stages and not at specified times, for the sake of convenience and simplicity, embryonic development is usually measured in hours or days. The photos below show the typical day-by-day growth rate for a chicken egg.

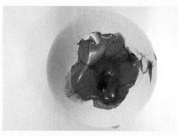

DAY ① The yolk sac begins to develop; the blastoderm grows to the size of a quarter and contains the start of the head (with brain and eyes), spine, blood vessels, digestive system, and nervous system

DAY ② The amnion and chorion begin to develop; faint blood vessels (the vitelline vessels) appear on the surface of the yolk, along with the start of the heart and ear; by the end of the day the heart (the spot in the center) begins to beat

DAY ③ The embryo, looking like a question mark, begins to take shape as the lungs, nose, legs, wings, tail, and allantois form and the amnion grows to surround the embryo; blood vessels become clearly visible

DAY ④ The embryo separates from the yolk sac, rises on the yolk, turns onto its left side, and begins to curve at the tail; the tongue begins to develop; the eye starts taking on color

DAY ⑤ Reproductive organs, gender differentiation, leg bones, crop, and circulatory system begin to develop; chorion and allantois begin fusing together; the start of a beak appears; the eye is now clearly visible

DAY ⑥ The three main segments of the legs and wings become distinct, the wings bend at elbow, and the legs bend at the knee; the chorioallantois develops at the blunt end of the egg; the embryo is now capable of voluntary movement

DAY ⑦ Feathers begin to form on the tail and thigh; the toes lengthen and separate, with thin webs between them; the comb begins to grow; the egg tooth starts developing; the chorioallantois attaches to the inner shell membrane

DAY ⑧ The neck and beak lengthen considerably; the upper and lower beak are of equal length; feather tracts begin to appear; the nictitating membrane (third eyelid) begins to develop; the skeletal bones start hardening

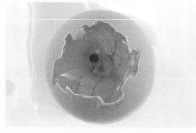

DAY ⑨ The beak continues to lengthen and a mouth opening develops; the upper eyelids begin covering the eyes; knee caps begin to form; the chorioallantois nears completion

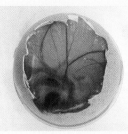

DAY 10 The toes start developing claws; flight feathers become prominent; the lower eyelid grows upward and the eyelids become more elliptical than round; the nostril narrows to a slit; the beak begins to harden

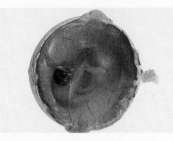

DAY 11 The toes curve and the legs develop scales; serrations appear on the comb; intestinal loops begin protruding into the yolk sac; the aorta (main blood vessel) is visible along the neck; the embryo becomes extremely active

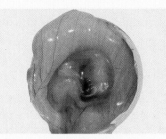

DAY 12 The toes begin to harden; the lower eyelid covers about two-thirds of the eye; most of the body is covered with the beginnings of feathers; fusion of the chorioallantois is complete and the membrane surrounds the embryo

DAY 13 The leg scales begin to overlap; down becomes clearly visible; the eyelids come together to form a crescent; the skeleton is nearly completely formed; the embryo produces increasing heat and uses increasing oxygen

DAY 14 The skull starts hardening; the head turns toward the blunt end of the egg; the growing embryo is fully formed; now as long as the egg is wide and getting crowded inside the shell, the embryo becomes less active

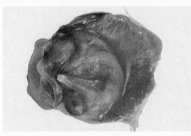

DAY 15 The outer covering of the beak has peeled away, making the beak appear shorter and blunter; the embryo completely fills shell expect for part of the air cell

DAY 16 The beak, turned toward air cell, sometimes presses against the cell membrane when the embryo rearranges its position; the intestinal loops (visible near the hock) start retracting into the body; the albumen is almost all absorbed

DAY 17 The embryo's body is nearly covered with down; it begins assuming hatching position — although the beak is typically beneath the right wing, other normal positions include beak over wing, feet over head, and head between thighs

DAY 18 The blood volume and hemoglobin (red blood protein) decrease, causing the embryo to appear somewhat pale; the embryo positions its body longwise to the shell in preparation for hatching

DAY 19 The intestinal loop finishes retracting and the yolk sac begins to enter the body and become enclosed within the abdomen; the beak breaks into air cell and the embryo takes its first real breath; the bird begins peeping

DAY 20 The yolk sac is half enclosed within the embryo's body; the blood content of the chorloallantoic membrane decreases and the membrane becomes sticky; the embryo completely fills shell except for part of the air cell

DAY 21 The neck begins to spasm and the embryo pips through the shell, then slowly rotates counterclockwise to break free; the shell here has been cut away to show the typical hatching position

Chicken Egg Development Compared to Eggs of Other Species

INCUBATED CHICKEN EGG	INCUBATED DUCK, GOOSE, TURKEY, AND GUINEA EGGS	MUSCOVIES, AND EGYPTIAN AND CANADA GEESE	PERCENTAGE DEVELOPED
5 days	7 days	9 days	25%
10½ days	14 days	18 days	50%
16 days	21 days	26 days	75%
21 days	28 days	35 days	100%

Egg Culling

Assuming you incubate only properly stored, clean eggs, the majority of your fertile eggs should hatch. The average hatching rate for artificial incubation is 85 percent of fertiles. The majority of those that fail to hatch stop developing and die at two peak times. The first major embryo death loss occurs within a few days of the beginning of incubation. The second, usually larger death loss occurs just before the hatch.

Culling, or removing, the eggs that are infertile or contain dead embryos has two purposes: One is to free up incubator space for viable eggs. The other is to remove potential sources of contamination. The multiplication of bacteria within a rotting egg may cause the egg to explode, spreading bacteria throughout the incubator and contaminating other eggs.

The timely removal of nondeveloping eggs helps you avoid the unpleasant experience of a rotten-egg explosion. Two ways to examine incubating eggs for viability are by candling them and by monitoring heart rates.

CANDLING THE EGGS

To candle eggs you'll need a candling device as described on page 152. Use it to examine eggs twice during incubation. Candle the first time about 1 week after incubation began, and the second time just before the hatch. A handy time for the second candling is while removing the turning device or moving the eggs to a hatching tray or hatcher. Do not candle after turning has stopped; during this critical period the chicks are positioning themselves for the hatch.

During both sessions work as quickly as possible. If you have a lot of eggs to candle, remove them in batches to avoid keeping them out of the incubator for too long.

At the first candling you can expect to remove as much as 10 percent of all the chicken eggs set, and slightly more than that for other species. At the second candling you can expect to remove about twice as many eggs as during the first candling.

Candling eggs with shells that are thick, dark colored, or speckled can be particularly difficult. A good way to get the hang of candling is to incubate some eggs with pale or white shells and use them to practice your candling technique. Once you know what you're looking for, and at, you can transfer that knowledge to eggs that are more difficult to candle.

Another way to hone your candling technique is to break open culls to verify your findings. Take the eggs outside, in case the contents smell bad. Even after you gain confidence in your candling abilities, continue breaking and examining culls. Determining the probable cause of deaths (as discussed below and in chapter 10) and tracking your findings will help you see patterns that allow you to improve your future hatching success.

First Candling

When you examine eggs after about 1 week of incubation, you will likely see one of the following things:

- A web of blood vessels surrounding a dark spot — the embryo is developing properly; you might even see it move
- Nothing — a clear egg is either infertile or fertile but defective and therefore unable to develop (an egg that's been refrigerated will candle clear)
- A thin ring within the egg or around the short circumference — the developing embryo died early in incubation
- A vague, cloudy shadow — the embryo died after several hours of incubation

The most common cause of embryo deaths clustered during early incubation is improper egg handling or storage. Washing dirty eggs or incubating soiled eggs, storing eggs under less than optimal conditions, or storing them for too long prior to setting them can contribute to early embryonic deaths.

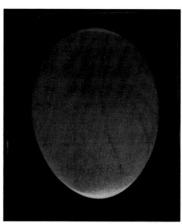

After 1 week of incubation, a properly developing embryo appears as a web of vessels surrounding a dark shadow.

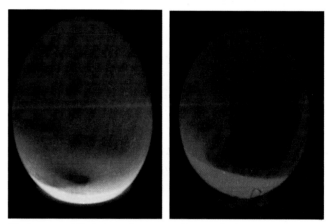

Shadowy, muddled, or murky contents in a candled egg are a clear indication that the embryo has died.

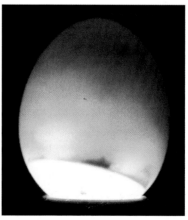

A blood ring appears as a thin, irregular circle around the shell's circumference, indicating the embryo has died.

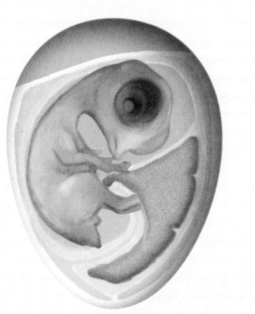

After about 2 weeks of incubation, the chick fills so much of the shell that candling reveals little more than an air cell and a dark form; movement of the dark form against the air cell tells you the embryo is alive.

Second Candling

When you examine eggs a second time, just before turning stops, you will likely see one of the following things:

- A dark shadow *in the air cell* (perhaps with movement against the air-cell membrane) — the embryo is developing properly
- Murky or muddled contents that move freely or a jagged-edged air cell — the embryo has died

Embryo deaths at this stage may be caused by the incubation of unclean eggs or by improper incubation procedures. They may also result from an inadequate breeder-flock diet. Hatchability issues are discussed in detail in chapter 10.

This late in incubation the still, dark mass of a dead embryo can look awfully similar to the dark form of a developing embryo that becomes less active as it runs out of room to move. In such a case sometimes the only way to tell the difference is to watch for movement against the air cell, a process that may lengthen the amount of time the eggs are out of the incubator at a critical time, thereby stressing the embryos. A swift but sure way to examine eggs at this stage is with a digital egg monitor.

MONITORING HEART RATE

A digital egg monitor is somewhat pricier than an egg candler, but it eliminates the guesswork, especially when candling eggs with thick, dark, or speckled shells and eggs containing late-term embryos. An egg monitor uses infrared technology to detect and display on a readout screen the heart rate of the embryo inside an egg placed on a sensor pad.

The heart rate of a developing embryo is incredibly fast. In contrast to the 72 beats per minute of the average human, an embryo's heart rate is in the range of 250 beats per minute. If you leave an egg on the monitor for a minute or two to cool down, you would see the heart rate slowing, as would happen when a setting hen leaves her nest for a daily bite to eat.

An embryo's heart rate is highly variable, changing from minute to minute as well as fluctuating over the entire incubation period. It tends to increase about two-thirds into incubation, decrease just before pipping activity begins (the embryo is resting up for the major push to get out of the shell), and increase while the embryo breaks into the air cell and breathes with its lungs for the

A digital egg monitor records the heart rate of an embryo in an egg placed on the sensor pad.

first time. Then follows a rest period during which the heart beats at a slower rate, speeding up again when the embryo pips through the shell and slowing down during rest periods.

Interestingly, the heart rate of male embryos on average tends to be slightly lower than that of females. Unfortunately, the rate among individuals is too variable to allow this criterion to be reliable for sexing eggs before they hatch.

Heart rate increases significantly after the hatch. The heart rate of a newly hatched chick, depending on its breed and gender, ranges between 350 and 450 beats per minute.

Average Heart Rate of Developing Embryo*

	CHICKEN		TURKEY		DUCK		GOOSE	
Stage	Day	Bpm	Day	Bpm	Day	Bpm	Day	Bpm
60%	13	255–271	16–17	227–245	16–18	245–271	17–18	243–255
70%	15	269–285	19–20	230–250	19–21	235–267	20–21	224–250
80%	17	278–296	22–23	236–256	22–24	234–260	23–24	207–241
90%	19	261–285	24–26	220–238	**	**	26–27	227–261
Pre-pip	19–20	252–284	25–27	219–241	**	**	Varied	217–255

*In beats per minute (bpm)
**Heart rate immeasurable because of excessive activity as hatch approached

Adapted from: "Embryonic Heart Rate during Development of Domesticated Birds," H. Tazawa et al., *Physiological Zoology*

The Hatch

Precisely when an egg will hatch depends on a lot of things including species, breed, strain, conditions under which the egg was stored prior to incubation, and conditions during incubation. Eggs that were held for a while before being set, or were set in an incubator that runs on the slightly cool side, will take longer than usual to hatch. If the incubator is run slightly on the warm side, the eggs will hatch sooner than usual.

In general, large eggs — such as those of the Jersey Giant — take longer than usual by as much as 2 days, while smaller eggs — including most bantams — tend to hatch a day or two early. Eggs of the Serama, the smallest of all bantams, may hatch in 17 days or even fewer. The accompanying table gives average incubation periods for various species of backyard poultry.

Because different species and breeds have different incubation periods, when combining eggs that you want to hatch all at the same time, schedule your settings accordingly. For instance, bantams and White Leghorns typically hatch a day early, so when combining their eggs with those of other breeds, add them to the incubator a day later than the other breeds. Likewise, bantam duck and Campbell eggs have a shorter incubation period than other breeds, so when hatching them along with other duck eggs, set them a day later. If you hatch both chickens and ducks, begin incubating the duck eggs a week before the chicken eggs.

Incubation Periods

SPECIES	DAYS
Bantam	19–20
Chicken	21
Duck (most breeds)	26–28
Duck, Muscovy	33–37
Guinea	26–28
Goose (most breeds)	28–32
Goose, Canadian and Egyptian	33–37
Turkey	26–28

HATCHING CONDITIONS

The optimal temperature for hatching is 0.5 to 1°F cooler (0.3–0.5°C) than that for incubation, with humidity 6 to 10 percent higher. These conditions may be achieved easily when all the eggs in the incubator are set to hatch at the same time. In fact, your hatching rate may be improved by providing higher humidity early in incubation and reducing it as the hatch approaches, which may be achieved by closing vents for the first week for chicken eggs (2 weeks for others), then opening them. During early incubation, embryos require less oxygen and generate less carbon dioxide than is the case by the time the vents are opened. During the hatch, restrict ventilation to raise humidity and prevent shell membranes from drying out. When about half the eggs have hatched, increase ventilation and reduce the temperature by about a degree to adjust for heat generated by hatching activity.

When eggs are placed in the incubator at different times and therefore hatch at different times, conditions must be averaged to accommodate eggs at the various stages of incubation. Using a separate hatcher is an alternative.

A still-air incubator makes a fine hatcher for small numbers of eggs and may be used to hatch eggs moved either from a larger incubator or from under a hen when the hatchlings might need more protection than the hen can provide. Move the eggs to the hatcher 3 days before they are scheduled to hatch. When about half the eggs have hatched, increase ventilation and reduce the temperature by a degree. Avoid opening the incubator during the hatch, as doing so causes the humidity to drop and reduces the rate of hatch.

PIPPING

About 3 days before the hatch is complete, pipping will start. Pipping is the act of breaking through the shell. Internal pipping occurs when the embryo breaks into the air cell and begins to breathe. External pipping occurs when the embryo breaks through the shell. By this time the embryo has drawn calcium and other minerals from the shell, causing the shell to become brittle and making it easier to break through.

The embryo pips with the help of its egg tooth — a small, sharp temporary cap at the tip of its upper beak. As hatching time approaches, and the embryo can no longer get enough oxygen through the shell pores, it uses the egg tooth to break into the air cell at the blunt end of the egg. There it finds sufficient oxygen to give it time to break out of the shell. A pipping muscle at the back of the tiny bird's neck begins to spasm, giving the bird enough impetus to pip through the shell's outer membrane, then through the shell itself to make its escape.

A normal chicken hatch is complete within 24 hours of the first pip. Larger species may take as long as 48 hours. When a bird cracks the shell all around the short circumference, and humidity is adequate to prevent the membrane from drying out, the hatch will probably go off without a hitch. But if the bird pips in one spot and doesn't soon continue around the shell, it is probably stuck, either because it is not properly positioned to turn its head or the membrane dried out too fast.

When the hatch is short and quick, move the hatchlings to a brooder when most of them (95 percent) have dried and fluffed out, leaving the wet ones in the incubator a little longer. Moving chicks while they're still wet can cause them to chill and die. If the hatch is slow (see Draggy Hatch on page 190), the early hatchlings will be okay in the hatcher for up to 24 hours after the hatch, after which they run the risk of dehydration and must be moved to the brooder. Avoid frequently opening the incubator to remove early hatchers, and work quickly while removing them, to avoid reducing the temperature and humidity and thus reducing the percentage of remaining eggs that will hatch.

A CHICK HATCHES

When a chick is ready to hatch, it uses its egg tooth to pip through the shell and shell membrane, then rests for up to 8 hours.

Again using its egg tooth, the chick rotates counterclockwise while chipping thousands of times to break through the shell about three-quarters of the way around, which can take up to 5 hours.

As the chick makes its way around the shell it keeps pushing against the shell cap, trying to straighten its neck and break free.

After about 40 minutes of hard work, the chick gets free of the cap, then takes a short rest.

With one mighty kick, the chick pushes free of the shell.

Wet and exhausted from its prodigious effort to enter the world, the chick sleeps.

Dispose of Incubated Eggs and Shells

Do not feed your hens unhatched eggs or the shells from incubated eggs. They're loaded with bacteria and can spread disease.

Within a few hours the chick's down dries; soon, no longer needed, the egg tooth will dry up and fall off.

HELP-OUTS NOT RECOMMENDED

A help-out is a hatchling that is unable to get free of the shell without assistance. The following conditions lead to help-outs:

Improper positioning. The bird is not properly positioned for hatching. Just prior to hatching, a bird should have its head under its right wing and be in a position to break through the air cell. A chick with its head under its left wing or between its legs cannot turn its head to break through the shell. Improper position can result from high incubation humidity, failing to keep the blunt end of the egg higher than the pointed end, or failing to stop turning the egg in time for the bird to reposition.

Air cell ill-positioned. The air cell is not at the blunt end of the egg, which can be caused by improper turning or by positioning eggs on the hatching tray with the pointed end higher than the blunt end.

Poor humidity. The hatching humidity is too low, causing the emerging bird to bind to the shell membrane.

When the blunt ends of the eggs remain above the pointed ends throughout incubation, and the temperature and humidity are properly controlled, help-outs could be a hereditary issue. In that case, birds that need help breaking free of the shell, assuming they live, will mature to produce more chicks that have difficulty hatching. For this reason — and because help-outs nearly always have crooked feet or a twisted neck that makes walking and eating difficult — assisting help-outs is not a good idea, especially if the chicks are destined to become breeders.

However, some people just can't resist. If that applies to you, do not lend assistance unless a shell has been pipped. If a bird gets stuck after breaking the shell in a circle, your timely assistance might save it. If it pips the shell but does not start circling within 12 hours, gently chip the shell from around the pip hole. If you see blood, stop and wait a few hours to avoid hemorrhage. And remember, if you keep opening the incubator to see what's going on, you contribute to the problem. The trick to successful help-outs is knowing when to lend a hand and when to let nature run its course.

Cleanup

An essential part of using an incubator is cleaning it after the hatch. The heat and humidity within an incubator offer ideal conditions for germs to flourish, and hatching produces plenty of organic wastes for the germs to thrive on. A buildup of hatching debris will eventually lead to a poor hatch rate.

No incubator is easy to clean and sanitize, but some are easier than others. A Styrofoam incubator is the most difficult type to clean. My experience, and that of others, is that after a few consecutive hatches the success rate drops sharply. Although a Styrofoam sanitizer is on the market, some incubator manufacturers recommend cleaning Styrofoam only with plain water. Easily sanitized plastic liners are available for

Cleaning Tip

If your incubator has a debris tray beneath the hatching tray, cleanup will be quicker and easier if you line the debris tray with aluminum foil. After the hatch, carefully roll up and dispose of the foil, along with most of the mess resulting from the hatch.

Foil may be used to line the bottom of other incubator models, provided it doesn't prevent evaporation from the water pan. If you have a Styrofoam incubator with an optional plastic liner, pulling it out and scrubbing it off likewise simplifies cleanup.

some models. As an alternative you can press aluminum foil against the bottom, shiny side down, and poke holes for the necessary air vents. If you opt for a Styrofoam incubator, plan on one hatch, or only two or three consecutive hatches, then thoroughly clean the incubator, and let it sit for several months before using it again.

THE CLEAN ROUTINE

The best sanitary measure for any incubator is to hatch one setting of eggs, then immediately clean and disinfect the incubator. As soon as the hatch is over — and all the hatchlings, shells, and unhatched eggs have been removed — unplug the incubator, remove all movable parts, and wash all those that are washable. Vacuum away dry debris, then lightly brush off the fan, heater, and thermostat with a small soft-bristle paintbrush, followed by gently passing a vacuum hose over them to suck out fluff. An old toothbrush comes in handy for cleaning corners and hard-to-reach areas, such as along tray runners. Do not attempt to clean your incubator controls with pressurized air or any other pressure device, as it could force debris into the controls and cause serious damage.

After the incubator is as clean as you can get it, scrub it with warm, soapy water. Take care not to wet down or spray the heater or any electrical parts. If you are done hatching for the year, all you need to do is leave the incubator open until it dries thoroughly — preferably in sunlight — then store it in a clean place. Any bacteria that might have survived the cleaning, scrubbing, and drying are unlikely to survive a lengthy period in a clean, dry storage area.

But if your hatching season will continue, follow up with scrubbing with a good disinfectant. Don't be tempted to skip the cleaning and scrubbing, because even the best disinfectant is ineffectual against organic debris.

Disinfect

Quaternary ammonium compounds (such as Roccal) are among the best disinfectants for incubators. They are sold in many places for all sorts of uses, from cleaning dog kennels to sterilizing surgeons' hands, and may be found under a variety of brand names. They leave no stain and have no odor, yet are strong and effective when used according to directions. Other sanitizers approved for use with poultry are phenol based (such as Tek-Trol), iodine based (such as Vanodine), or chlorine based (such as Oxine). An inexpensive alternative is chlorine bleach diluted ¼ cup to 1 gallon (15 ml/l) of hot water.

All disinfectants are hazardous to one extent or another. A green alternative is to use plain 5 percent vinegar. Distilled white vinegar and apple cider vinegar work equally well for this purpose.

Apply the sanitizer with a spray bottle or clean sponge or washrag, avoiding controls and metal parts. Pay closest attention to areas where hatching debris collects, including the hatching tray and the bottom of the incubator. Wait 15 minutes, then wipe down the incubator with a fresh sponge or clean cloth and warm water. Thoroughly dry the incubator, preferably in sunlight, or leave the water pan empty and run the incubator for a day or so until it dries.

Ideally, an incubator or a hatcher should be thoroughly cleaned after every hatch. If your first hatch is successful, but your subsequent hatch rate goes downhill, you can be pretty sure your incubator was not properly cleaned and disinfected. Cleaning after every hatch is no problem if you use a separate hatcher or a small tabletop incubator in which all the eggs hatch at once. Cleaning does become problematic when batches of eggs are added periodically and set to hatch on different days (typically a week apart) in the same unit in which they are incubated. In such a case you'll be happy to have an incubator that is designed for ease of cleaning.

WHAT WENT WRONG?

Any number of things can go wrong between the time you put eggs into your incubator and the time they should hatch. Some things (such as a power outage) may be beyond your control. Some, largely due to carelessness, may be of your own making. But through diligent observation, you can determine many of the factors that affect hatchability and correct them in the future for greater hatching success.

Record Keeping

Incubators and their owner's manuals are designed to average out conditions for hatching any type of poultry. But optimal conditions differ for different species, breeds, and strains within a breed. By keeping detailed records for each hatch, you can fine-tune your incubation technique to vastly improve your future rate of hatch. Important information to record includes the following:

- Source of the eggs (hen, flock, breed, strain, and so on)
- Date and time eggs are set
- Number of eggs set
- Date and results of first candling
- Date and results of second candling
- Date turning should stop
- Date hatch should start
- Date hatch actually started
- Date hatch completed

Likewise, record any events that may affect the hatch. You might make these notes on a calendar so you can track exactly when during incubation each event occurred and how long it lasted. These notes might include such things as the water pan having run dry, deviations in incubation temperature, power outages, and changes in weather conditions, especially rain and other high humidity levels or extremely dry weather.

As incubation proceeds you might recognize where future changes in your procedure might be beneficial; jot those down, too. After the hatch, some of the hatchlings may not make it if they are too weak or deformed to survive past the period when yolk reserves can continue to carry them. Note those, as well, especially if they occur after every hatch.

The more detailed your records, the easier you'll be able to spot patterns and find room for improvement. Detailed hatching records will not only help you fine-tune your incubation procedures in the future but will help you identify breeder-flock management issues, such as poor nutrition or inbreeding depression, that may affect the hatch.

Talking Fowl

asexual reproduction. Also called *parthenogenesis;* the phenomenon whereby an unfertilized egg develops an embryo

blood ring. The appearance of a candled egg in which the embryo has died early, causing blood to accumulate in a ring circling the egg's short circumference

breakout analysis. The examination of an incubated or partially incubated egg, typically to determine why the embryo failed to develop properly

crest. A spherical puff of feathers growing on the head of a breed that has a knob on top of its skull

exploder. A rotting egg that literally explodes in an incubator or under a hen

gynandromorph. An individual having both male and female characteristics and organs

lethal genes. A gene that can cause death, typically during incubation

malposition. Any position of a developed embryo that would prevent it from being able to free itself from the shell at the time of hatch

rumpless. Lacking a tailbone and therefore lacking a tail

zygote. A fertilized egg before embryonic cell division begins to take place

Breakout Analysis

By breaking open and examining eggs that are culled during candling or that fail to hatch, you can often find important clues as to what went wrong. Using a digital camera to document your findings will help with your record keeping and also let you identify and compare eggs with similar problems, whether from the same batch or from two or more settings.

I take eggs outdoors for breakout. Some of them don't smell real great, and occasionally, one will pop when I tap the shell. To contain the mess I use a large flat container, such as a dishpan or clean cat litter box. I break the eggs into one end, and after examining the contents, I use a gardening hand hoe to scrape them to the far end before breaking the next egg. The breakout procedure is as follows:

- If the shell has pipped, before breaking into the shell, note the location of the pip in relation to the air cell.
- Gently crack into the shell at the blunt end and peel away the shell membranes.
- If it is a term embryo, check to see if it pipped into the air cell.
- Note the embryo's position.
- Gently peel the shell away from the membrane enclosing the embryo.

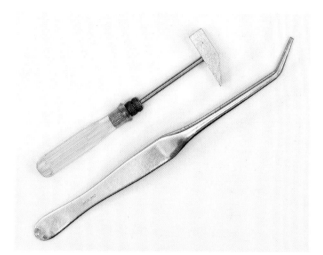

Handy tools for break-out analysis are a miniature hammer for cracking the shell and a pair of tweezers for peeling away the membranes.

- Note any deformities or anything else unusual.
- If the dead embryo is non term, try to determine the age at death (Stages of Incubation on page 170 may prove helpful here).
- If the egg candled as clear, look for signs of fertility — blastodisc, blastoderm, or early-death embryo.

Track Your Hatch Rate

LINE	ITEM	NUMBER
A	Eggs set	
B	Not fertile	
C	Dead at first candling	
D	Dead at second candling	
E	Full-term, failed to pip	
F	Pipped, failed to hatch	
G	Total lost = B + C + D + E + F	
H	Total hatched = A − G	
I	Percent hatched = H × 100 / A	%
J	Percent of fertiles hatched = H × 100 / (A − B)	%

The first few times you break eggs for analysis, you probably will not be entirely sure what you are supposed to look for. Just keep looking and making notes until you start to see patterns. The first few times you examine eggs that failed to hatch, use a simple list such as the one shown in Track Your Hatch Rate opposite. As you break out and examine each egg, make a mark next to the category it most closely falls into.

After you gain some experience breaking out and analyzing eggs that fail to hatch, you will begin to see details you missed as a newbie. You will then be better able to scan through the following possibilities, eliminate those that don't pertain, and come to some conclusions that will be helpful in fine-tuning your future hatches.

Just remember that not every egg will hatch. Occasional issues are not uncommon or particularly problematic, but pay attention to problems that affect several eggs in the same stage of development, as well as for problems that group into two or more stages of development that may all relate to a single cause.

CLEAR EGG

An egg that appears clear when candled is either infertile or fertile but failed to develop properly. If more than 10 percent of eggs are clear when candled after a week of incubation, learning to tell the difference between fertiles and infertiles becomes an important troubleshooting skill that will help you determine if the problem lies with your breeder flock or your incubation technique.

Infertile

An incubated infertile egg has a germinal disc, as shown in the photo on page 142, with no sign of blood. Not all eggs will be fertile, but a high rate of infertility is a bad sign. Infertility has many causes. One of them, of course, would be no male in the flock. Another cause may be an incorrect ratio of hens to males. When a male is expected to service too many hens, he may not be able to get around to them all or may run out of sperm before he does. When a flock includes too many males, they may spend too much time fighting among themselves and have little time left to get around to the hens, or they may interfere with each other's attempt to mate.

Even if the mating ratio is optimal, fertility will be low in a breeding flock that's too closely confined; housing that's too small for the number of breeders can result in reduced mating frequency. In a small flock with only one male, he may prefer some hens and ignore the others. Hens that are high in the peck order tend to be mated less often than hens that are lower in peck order because they expect to be pecked when another bird approaches and thus make easy targets as mates.

Age can affect fertility. Infertility may result if breeders are not sexually mature enough to produce viable sperm (males) or eggs (females). Infertility may also result from breeders that are too old, especially males. For most species fertility declines sharply after about the third year, and a five-year-old male may no longer be able to cut it.

A hen that's too fat or too thin or a male that's too fat may have fertility problems. A male with a leg or foot injury may have trouble breeding. Excessive exhibition of breeders — males or females — can cause stress that leads to infertility.

Nutrition can affect fertility, as can disease; both these issues are discussed more fully later in this chapter. Parasite problems, internal or external, can cause enough discomfort to interfere with fertility by reducing frequency of mating. Exposure to pesticides and other toxic chemicals, or to some medications, can reduce fertility.

Extremely hot or cold weather can affect fertility, which tends to be greatest at temperatures between 55°F and 80°F (13–27°C). Season can affect fertility, which tends to be low during the times of year when daylight hours are fewer than 14. Just as lighting keeps hens laying during short winter days, it also improves the fertility of cocks. Fertility increases from December to April, is highest in spring, and then rapidly declines during the heat of summer. The first egg laid in a clutch may be infertile, as may those laid during a molt.

Breed-related mechanical problems may also result in low fertility. Such mechanical problems include these:

Comb size. Breeds with large single combs have trouble negotiating feeders with narrow openings, and the resulting nutritional deficiency affects fertility. Large combs are more subject to frostbite, and frozen combs and wattles can cause sufficient pain to deter a male from mating. The same is true of geese with frozen knobs.

Crests. Houdans, Polish, and other heavily crested cocks may not see well enough to catch hens. For a quick fix, clip back their crest feathers.

Heavy feathering. Brahmas, Cochins, Wyandottes, and other heavily feathered breeds have trouble mating. Fertility may be improved by clipping vent feathers.

Foot feathering. Booted bantam cocks and males of other breeds with heavy foot feathering have trouble getting a foothold when treading hens.

Rumplessness. Araucanas (and occasionally birds of other breeds) have no tail to pull the feathers away from their vents during mating. The quick fix is to clip the vent feathers of both cocks and hens, with more attention to the feathers above the hens' vents and those below the cocks' vents. A better solution is to select breeders with the least vent feathering.

Heavy muscling. Cornish cocks, broad-breasted turkeys, and other heavy-breasted males have trouble mounting hens because of the wide distance between their legs. These breeds are typically bred through artificial insemination.

Fertility issues may be genetically related. Cocks that sport rose combs generally have lower fertility than single-comb cocks. Any breed, or strain within a breed, that is highly inbred because of a low population also tends to be low in fertility.

Fertile, No Embryo

A fertilized egg, or zygote, contains all the essential elements for the creation of a baby bird, but they remain encoded as a set of instructions until incubation allows an embryo to begin developing. A zygote may die between the time the egg leaves a hen's ovary and the time incubation begins — a phenomenon sometimes known as weak fertility. An incubated fertile egg containing an enlarged blastoderm but no blood indicates that embryo development did not occur. Causes of fertility without an embryo include the following:

- Eggs left in the nests too long before being collected for hatching
- Eggs were roughly handled, such as being shaken during collection or transport
- Soiled eggs were washed in too-hot water
- Washed eggs were improperly sanitized
- Eggs were stored too long prior to incubation; see Egg Storage on page 154.
- Eggs were stored under improper conditions prior to incubation
- Eggs were subjected to temperature shock
- Incubator temperature started out too high
- Breeders are too young or too old
- Breeders are too inbred
- Breeders are diseased
- Breeders have been medicated
- Breeders have hereditary issues
- Breeders or eggs were exposed to pesticides or other toxic chemicals

Asexual Embryo Development

Asexual reproduction, or parthenogenesis, is a little-studied phenomenon that occurs in turkeys, chickens, and most likely other poultry as well, whereby an unfertilized egg develops an embryo. No one knows exactly how common it is, because most parthenogenic embryos die at the start of incubation and may be mistaken for a dead fertilized embryo.

Asexual reproduction apparently is more likely to occur in some breeds and strains than in others. It has been identified in Cornish, Leghorn, and Plymouth Rock chickens, as well as in different strains of white turkeys. Some of the conditions under which it may occur include the following:

- Eggs from young hens
- First eggs in a clutch
- Double-yolk eggs
- Eggs stored at too warm a temperature
- Eggs laid by diseased hens or those having been vaccinated with a live virus
- Breeders fed certain additives, including yeasts such as those found in probiotics

Four categories of asexual reproduction have been identified in incubated eggs, with the first two being more likely than the last two: unorganized tissues; ruptured yolk sac membrane, causing the yolk's contents to mix with albumen; blood ring or blood spots; developing embryo that, in extremely rare cases, may hatch. When one of these eggs does hatch, incubation takes 2 days longer than for a normal embryo. Images of turkey eggs showing all four categories may be found at http://oregonstate.edu/instruct/ans-tparth.

Asexual reproduction looks so much like normal early embryonic death that embryo deaths attributed to the first eggs in a clutch and to eggs laid by young hens may actually be asexual development, considering that young hens lay in shorter clutches than older hens and therefore lay a higher percentage of first eggs. The obvious way to verify asexual reproduction, also known as virgin birth, is to incubate eggs from hens that have not been with a male. However, should you be so lucky as to get one to hatch, preliminary indications are that it would most likely be a male.

Possible indications of asexual reproduction are rupturing of the yolk sac membrane, allowing the yolk to mix with the albumen, and embryonic blood cells appearing as a blood ring.

BLOOD RING

A fertile egg that began to develop an embryo but died during the first 3 days of incubation (for chickens, 4 days for other species) has either a blood ring with no apparent embryo or a tiny developing embryo that died before the eye became visible. Eggs in this category often appear along with a large percentage of infertile eggs. Causes include the following:

- Eggs were stored too long prior to incubation; see Egg Storage on page 154
- Eggs were stored under improper conditions prior to incubation
- Incubator temperature started out too high or too low
- Eggs were roughly handled during collection or transport
- Eggs experienced temperature shock prior to incubation
- Breeders are too old
- Breeders are too highly inbred
- Breeder diet is seriously deficient
- Breeders are diseased
- Breeders or eggs were exposed to pesticides or other toxic chemicals
- First egg laid at the beginning of a new clutch

EARLY DEATH

An embryo that died between 3 and 6 days into the incubation period (4 to 8 days for species other than chicken) has a developed yolk sac system, is lying on its left side, has a visible eye, but does not yet have an egg tooth. All the factors that result in a clear fertile egg with embryo can cause early embryo death as well. Other causes include:

- Insufficient ventilation, causing too high a concentration of carbon dioxide
- Insufficient turning
- Improper turning angle
- Breeder-flock diet is deficient in vitamins
- Mold or other incubator contamination

MID-TERM DEATH

Typically, the fewest embryo deaths occur during midterm, which lies between days 7 and 17 for chickens, days 9 and 23 for other species. During this period the embryo develops claws and an egg tooth, and toward the end of this period, feathers begin to appear. Causes of midterm death include:

- Improper operation of the incubator, which could involve temperature, humidity, turning, ventilation, or all four parameters
- Incubator contamination
- A nutritionally deficient breeder-flock diet
- Lethal genes, as discussed on page 203.

Exploding Eggs

If an egg is going to explode, it usually does so about 12 days into the incubation of chicken eggs, 16 days for others. Whenever you open your incubator and smell an unpleasant odor, take time to look for and remove the offending egg or eggs.

The normal odor of an egg is governed by a number of volatile compounds that include sulfides in the form of small amounts of dimethyl sulfide, dimethyl trisulfide, and hydrogen sulfide. When bacteria penetrate a shell and break down egg protein, they produce excessive amounts of hydrogen sulfide, which results in the familiar odor commonly known as the rotten egg smell.

A rotting egg may or may not exude darkish fluid that beads on the shell or (due to the pressure of gases within the shell) may crack and leak. Touching such an egg can cause it to explode in your hand, contaminating other eggs in the incubator.

Exploders result from incubating dirty or contaminated eggs. Eggs may get dirty in the nest or from being laid on the coop floor. They may become contaminated by being handled with dirty hands, washed in water that is cooler than the egg, rinsed in contaminated water, stored while still damp from being washed, or stored in contaminated cartons. Contamination may occur during incubation, resulting from setting eggs with soiled, cracked, or leaking shells, or from eggs that previously exploded during incubation.

Eggs can become contaminated through sweating, which occurs when eggs are moved from a cool to a warm (and especially humid) environment, causing moisture from the air to condense on the shells. This moisture promotes the growth of bacteria, which then may penetrate through pores in the shell. Sweating can cause not only exploders but also dead embryos during incubation and the infection and deaths of hatchlings.

When removing a potential exploder from the incubator, wear plastic gloves. In case the egg should explode while you're removing it, cradle it in your hands to contain the mess or wrap it in a paper towel before picking it up.

Contamination multiplies rapidly when an egg explodes in the incubator. The only sensible thing to do is to empty the incubator, dispose of all eggs, clean and sanitize the incubator, and start over.

FAILURE TO HATCH

One of the saddest things that occur during incubation is having embryos die just before they should hatch. Some breeds are notoriously hard to hatch. Sebrights, for instance, are difficult to hatch, and the chicks generally lack hardiness. Bantams tend to lay eggs that are round in shape, making it difficult to determine at which end the air cell is located. Nearly every batch of eggs has a few genetically abnormal embryos, but having a large number of embryos reaching full term without hatching is a bad sign.

Most of the reasons for midterm embryo death are also related to late-stage deaths. General causes for late-term embryo death include:

- Eggs stored too long prior to incubation
- Eggs chilled during transfer from incubator to hatcher
- Eggs turned too long or transferred to hatcher too late
- Incubator or hatcher opened too often during the hatch
- Embryo improperly positioned for hatching (malpositioning)
- Twinning within a double-yolk egg
- Poor shell quality, which may result from age, disease, or dietary deficiency
- Eggs contain insufficient nutrients due to poor breeder-flock diet
- Breeder flock is diseased
- Genetic abnormality

Whether embryos die before or after pipping can give you additional clues.

Failure to Pip

A full-term embryo that fails to pip may still have a large yolk sac or an enlarged abdomen that has not fully closed. Causes of pipping failure include:

- Inadequate turning, resulting in decreased embryonic membrane development and poor nutrient absorption
- Humidity too high during incubation or hatching
- Temperature too low during incubation
- Temperature too high during the hatch
- Ventilation inadequate during the hatch

Dead in Shell

Dead in shell is the term used to describe a fully formed embryo that pips but dies after being unable to free itself from the shell. Causes for dead-in-shell embryos include:

- Humidity too low for long periods during incubation
- Humidity too low during the hatch
- Temperature too low for long periods during incubation
- Temperature too high during the hatch
- Poor ventilation during incubation or hatching
- Inadequate turning during the first 12 days of incubation
- Jarring of eggs during transfer to hatcher
- Opening incubator too often during the hatch

The majority of deaths in this category occur due to exhaustion or lack of oxygen. An embryo that pips and remains alive without being able to exit from the shell is essentially a dead-in-shell embryo that hasn't yet died; see Help-Outs Not Recommended on page 178.

When an Embryo Is Malpositioned

Failure to hatch often occurs because an embryo is malpositioned, meaning for some reason it did not move into proper hatching position during the last week of incubation. The normal position for an embryo during the last 2 days of incubation is to be oriented longwise within the egg, with its feet situated toward its head. The head should be at the blunt end of the egg, turned to the right, and tucked under the right wing with the beak pointed toward the air cell.

Possible malpositions, starting with the most common ones, include beak above right wing; feet over head; head between thighs; head under left wing; head at small end of egg; head not oriented toward air cell. Some variations are lethal; others are not.

Less than 2 percent of incubated eggs commonly fail to hatch because of malpositioning. Any larger percentage indicates a problem, which may include:

- Incubating round eggs
- Incubating larger than normal eggs for the breed
- Incubating eggs with the pointed end upward
- Improper incubator operation (temperature, humidity, ventilation), resulting in inadequate air cell development
- Temperature too low during the hatch
- Breeder hens are old, resulting in poor shell quality
- Breeder-flock diet is nutritionally deficient, especially in vitamins A and B_{12}

Troubleshooting the Hatch

Once hatchlings begin to appear, the way in which the hatch proceeds, as well as the way the hatchlings look and act, can give you more clues about your incubation technique and your breeder-flock management practices.

EARLY HATCH

Birds that hatch ahead of schedule tend to be thin and noisy. Causes for early hatching include:

- Incubation of small eggs
- Difference in incubation period among breeds or strains
- Temperature too high during incubation
- Humidity too low during incubation

When premature hatchlings have bloody navels, the most likely cause is too high a temperature during incubation or hatching.

LATE HATCH

Healthy embryos that are incubated under ideal conditions generally hatch right on schedule, or sometimes a little ahead of schedule. When a hatch is terribly late, it's tempting to think all is lost, but that's not necessarily the case. I once had a series of power issues delay a hatch by several days; it was not one of my better hatches, but under the circumstances a remarkable number of chicks and keets made it into the world. Causes for late hatching include:

- Incubation of large eggs
- Eggs stored too long prior to incubation
- Temperature too low during incubation
- Humidity too high during incubation
- Weak embryos, which could have a variety of causes including nutritional
- Breeder flock too old
- Breeder flock too highly inbred

Incubation Troubleshooting at a Glance

SIGN	CLEAR	BLOOD RING	STAGE AT EMBRYO DEATH		
			EARLY	MID-TERM	AT HATCH
Infertile	*		*		
Eggs left in the nests too long before collection	*				
Soiled eggs washed in too-hot water	*				
Washed eggs improperly sanitized	*				
Eggs stored too long	*	*	*		*
Eggs improperly stored	*	*	*		
Eggs experienced temperature shock	*	*			
Incubator temperature started out too high	*	*	*		
Incubator temperature started out too low		*	*		
Breeders too young	*				
Breeders too old	*	*	*		
Breeders too inbred	*	*	*		
Breeders are diseased	*	*	*		*
Breeders were medicated	*				
Breeders have hereditary issues	*				
Breeders or eggs exposed to toxic chemicals	*	*	*		
Eggs were roughly handled		*	*		*
Breeder diet is seriously deficient		*	*	*	*
First egg laid at the beginning of a new clutch		*			
Asexual reproduction	*	*			
Incubator contamination			*	*	*
Insufficient ventilation			*	*	*
Insufficient turning			*	*	*
Improper turning angle			*		
Incubation temperature too high				*	*
Incubation temperature too low				*	*
Incubation humidity too low				*	*
Incubation humidity too high				*	*
Lethal genes				*	*
Eggs chilled during transfer to hatcher					*
Eggs transferred to hatcher too late					*
Incubator or hatcher opened too often					*
Malpositioning					*
Hatcher temperature too high					*
Twinning in a double-yolk egg					*
Poor shell quality					*
Genetic abnormality					*

DRAGGY HATCH

Most hatches take 24 to 48 hours between the appearance of the first hatchling and the last to make it out of the shell. A draggy hatch, if caused by low temperature, may start late and usually will be accompanied by a large percentage of embryos failing to hatch. The most drawn-out hatch I ever experienced, which occurred due to a varying power supply at critical stages of incubation, dragged on for 4 days and yielded only about a 40 percent rate. Causes for draggy hatch include:

- Incubating small and large eggs together — larger eggs take longer to hatch
- Eggs stored for varying lengths of time prior to incubation — the eggs stored longer will take longer to hatch
- Setting eggs at the same time from breeds or strains that have slightly different incubation periods
- Incubating eggs together from both young and old breeders
- Temperature uneven throughout incubator
- Temperature too high or too low during incubation or hatching

DEAD IN INCUBATOR

Sometimes everything goes right during the incubation and hatching period, but when you open the incubator, you find some or all hatchlings dead. The most common causes of hatchlings dying in the incubator or hatcher are:

- Temperature too high during the hatch
- Insufficient ventilation, providing too little oxygen
- Hatchlings left in incubator so long they dehydrate

STICKY HATCHLINGS

Sticky hatchlings are covered with wet or dried goo, which is embryo waste. The most common reason for stickiness is too-low humidity, causing a bird to dry out before it can kick free of the shell and associated hatching residue. Eventually, the goo will slough off, but to help it along you can take matters in your own hands.

Remove the goo from a sticky chick by quickly rinsing it under lukewarm tap water, then dry the chick off thoroughly.

Working gently but quickly, rinse off the bird with warm tap water. Some folks submerge the bird up to its neck in a sink or bowl. I hold my sticky hatchlings under a faucet running a gentle stream of lukewarm water. Cleaning each bird should take no longer than it would to casually wash your hands.

Once the goo has washed away, wrap the baby in a clean absorbent towel. In my barn I keep boxes of old, clean towels and washcloths as well as a roll of paper towels in a holder for times like these. Sop up as much moisture as possible, remove the towel, then put the hatchling back into the incubator to keep it from chilling while it dries off.

Causes of sticky hatchlings include:
- Eggs stored too long prior to incubation
- Incubating eggs larger than normal for the breed
- Temperature too low during incubation
- Humidity too low during incubation or hatching
- Ventilation inadequate during the hatch

When Hatchlings Pant

If the hatchlings are panting in the incubator, the temperature may be too high. Panting releases heat through evaporation from the respiratory system. If the temperature is correct, the problem is most likely insufficient ventilation.

Improper ventilation during the hatch can cause the oxygen level to go down and the carbon dioxide level to go up. When that happens, baby birds have to breathe faster to get enough oxygen and release excess carbon dioxide. The simple solution to ensure hatchling comfort and survival is to improve ventilation.

SHELL CLINGS TO HATCHLINGS

Sometimes a hatchling pips all the way around but can't get free of the top or bottom part of the shell or has pieces of shell sticking to it after it dries out. Usually, the shell particles may be easily picked off. Small pieces of shell will eventually fall off without your help.

Causes for sticking shells may be any of the same factors that cause sticky hatchlings. Additional causes for clinging shells are humidity too low during egg storage and poor egg quality, both of which can result in dried-out shell membranes.

EYES STUCK SHUT

Hatchlings with their eyes stuck shut are most likely dehydrated. Causes include the following:
- Temperature too high during the hatch
- Humidity too low during the hatch
- Too much ventilation during the hatch
- Leaving hatchlings in the incubator too long after the hatch

Eyes may be unstuck with a drop of quality eyewash, such as Systane. If an eye remain stuck shut, dampen a Q-Tip with warm water and roll it (do not press or rub the eyes with the Q-Tip) across the eye to moisten and unstick it.

SMALL HATCHLINGS

Eggs hatched from a mixed-breed flock, especially if the breeds are of different sizes or include both large and bantam varieties, will produce hatchlings of varying sizes. Eggs hatched from a straightbred flock usually result in hatchlings of a fairly uniform size. Smaller-than-normal hatchlings are not a serious problem unless they are weak or have other issues aside from size. When a significant number of straightbred hatchlings are smaller than normal, causes include the following:

- Humidity too low during egg storage
- Humidity too low during incubation
- Incubating small eggs
- Temperature too high during incubation
- Incubating eggs with thin, porous shells
- Incubating at an altitude higher than 5,000 feet (1,500 m), which increases loss of water from the egg, slows the embryos' metabolic rate, and reduces their rate of growth

WEAK HATCHLINGS

Weak hatchlings are not as active as normal hatchlings, which usually start moving around as soon as they dry off. Weak hatchlings don't move much on the hatching tray, may not actively seek feed and water in the brooder, and therefore may die. Causes for weak hatchlings include:

- Temperature too high during the hatch
- Inadequate ventilation during the hatch
- Contamination in incubator or hatcher
- Breeder flock is in poor condition
- Breeder-flock diet is nutritionally deficient, especially in vitamins
- Breeder flock is diseased

UNHEALED NAVELS

Hatchlings with rough or bloody unhealed navels may either have dry, rough down or moist, soft, mushy bodies. Although the conditions are different, and generally have different causes, both conditions may be caused by too-high humidity during the hatch.

A hatchling with an unhealed navel may eventually heal, or it may die if the navel ruptures. When chicks smell bad, culling is the best resort, as they will likely die anyway.

Dry, Rough Down

Causes for unhealed navels coupled with dry, rough down include:

- Temperature too high during incubation
- Wide temperature fluctuations during incubation
- Temperature too low during the hatch
- Humidity too high during the hatch
- Breeder-flock diet is nutritionally deficient

Mushy Hatchlings

A mushy hatchling has a large, moist, soft body; has a swollen abdomen with an unhealed navel; and is typically lethargic. If it smells bad as well, it has the bacterial infection omphalitis, or mushy chick disease. Causes for mushy hatchlings include:

- Temperature too low during incubation
- Humidity too high during incubation or hatching
- Inadequate ventilation
- Contamination in incubator from dirty eggs or failure to clean and sanitize after previous hatches

CLUBBED DOWN AND CURLED TOES

Clubbed down is so called because each downy feather takes on the shape of a club, resulting from failure of the down sheaths to rupture, which bends each bit of fluff so it bulges out at the shaft base. Clubbed down is generally considered to be caused by a deficiency in riboflavin (vitamin B₂) and is more often seen in black breeds because the production of the black pigment melanin uses up riboflavin. However, clubbed down is sometimes associated with a normal hatch or a low percentage of dead-in-shells and is not necessarily a sign of riboflavin deficiency.

On the other hand, the combination of clubbed down with curled-toe paralysis is a good indication that the breeder-flock diet is deficient in riboflavin. In curled-toe paralysis the toes curve inward, causing the bird to step on its own toes. As a result of discomfort, the bird rests or attempts to move around on its hocks. While some affected birds survive, others starve because they can't get around to the feeder and drinker. Curled-toe paralysis is not the same as crooked toes, described on page 194. In the latter case, the bird walks on the sides of its toes but otherwise manages to move around adequately.

Riboflavin deficiency in hatching eggs may occur when the breeder diet consists solely of layer ration without vitamin supplementation or access to green forage. Riboflavin deficiency may also result from a poor-grade starter ration, in which case curled-toe paralysis typically doesn't appear until 10 to 14 days after the hatch.

Hatchlings with clubbed down can survive just as well as normal hatchlings, although they tend to chill more readily because of their abnormal feathering. Curled toes, on the other hand, become a problem if the bird is so deformed it can't adaquately get around to eat and drink.

A chick with the combination of clubbed down and curled toes is a fairly reliable indication that the breeder flock diet is deficient in riboflavin.

Live or Let Die? When to Cull a Sickly Chick

Old-time poultry keepers often point out that "cull" means "kill." But for some of today's poultry enthusiasts, cull is a four letter word of the worst kind, and "kill" isn't even in their vocabulary. I've seen folks go to great lengths to try to save the life of every hatchling, regardless of the cost (in time, money, and frustration) to the chicken keeper and the bird's inability ever to enjoy quality of life.

So how do you decide when to invoke the dreaded c-word? Here are a few guidelines:

- The bird is so physically impaired it cannot eat and drink properly. The bird may, for example, have a beak or bill deformity that keeps it from being able to obtain sufficient nourishment, or it may have a leg deformity that prevents it from adequately getting around to access food and water.
- The bird is being trampled or mercilessly picked on by brooder mates. Birds can sense when one among them is weak in some way and will make things worse for the weakling, or maybe even kill it.
- The bird is suffering. A bird that sits off by itself, looks depressed, and loses weight or fails to gain weight is what's known as "unthrifty" or a "poor doer." A bird that has to struggle to survive has no future and should not be made to suffer.
- If you are developing a breeder flock for exhibition purposes or to help to save a rare breed from extinction, you certainly don't want to introduce or perpetuate genetic weaknesses.

Robust hatchlings are a joy to raise. And when they mature they likely will produce more of the same. By not culling the occasional bird now, you may end up having no choice but to cull many more of its descendants in the future.

DEFORMITIES

The two most common hatchling deformities are crooked toes and splayed legs. Any number of other deformities, some of which are listed below, may be expected as occasional occurrences. However, their appearance in more than 5 percent of the hatchlings calls for investigation. A large number of deformities in the hatch often occurs in a poor hatch with a high percentage of dead-in-shells.

General causes of deformities include mishandling of hatching eggs during storage, especially jarring them; storing them under improper temperature and humidity; storing them for too long; or letting them chill before the start of incubation. Incubator operation can, of course, affect growth. A too-high temperature accelerates growth and can result in brain issues, as well as in one or both eyes missing. A too-low temperature, or cooling waterfowl eggs too long, retards growth.

Deformities may arise from breeder-flock issues that may be related to excessive age, to hereditary factors in the breed or strain, to disease, or to diet. An embryo obtains required nutrients from the entire egg: yolk, white, and shell. An insufficiency of protein, vitamins, and minerals in the breeder-flock diet is reflected in insufficient nutrients available to the embryo for proper growth, resulting in nutritional deformities.

Crooked Toes

Crooked toes may occur during incubation, during the hatch, or after the hatch and are more common in chicks and other land fowl than in waterfowl. They may result from high temperature early in incubation, low temperature throughout incubation or during the hatch, or excessive activity too soon after hatching. Brooder conditions associated with crooked toes are overcrowding and a too-smooth floor. Other causes include nutritional deficiency, injury, and heredity.

If you can't identify the cause, don't include crooked-toe birds in your breeder flock, or they may produce future crooked-toe generations. And a bird with crooked toes is not suitable for show.

Crooked toes sometimes may be caused by a too-low hatching temperature or from the chick's enjoying too much activity too soon after the hatch.

On the other hand, although crooked toes may not look great, a bird will get along fine in all other respects despite its bent toes.

Treatment. Crooked toes are most likely to be successfully straightened when the bird is fresh out of the incubator and its bones are still soft. Cut a piece of first-aid tape about twice the size needed to cover the foot. One-inch-wide Transpore first-aid tape works well for this purpose as it is sticks nicely, is moisture resistant, and has pores for breathability. Lightweight cardboard held to the foot with regular first-aid tape or even Band-Aids, if that's what's handy, also works.

Some people use duct tape, which works well to straighten toes but is difficult to remove and, if not removed carefully, can irritate the skin. Duct tape or other types of really sticky tape may be removed more easily with the help of baby oil, which in turn may be removed with an antibacterial hand sanitizer such as Purell. Such a sanitizer also helps disinfect areas that might have gotten irritated by the sticky tape.

Straighten the hatchling's toes and lay the bottoms against the sticky side of the tape (or against the cardboard), in normal position. Fold the tape

CROOKED TOE REPAIR

Crooked toes may be straightened by arranging the toes in their normal position and covering them with duct tape or first-aid tape.

over the top of the foot, creating a flipper-like shoe. The shoe should be no bigger than necessary to cover the foot, but should completely enclose the toes. If the toes are going to straighten out, they should do so within a day or two as the bones harden.

Splayed Leg

Like crooked toes, splayed or spraddled legs may occur during incubation, during the hatch, or after hatch. The condition may be caused by a too-high incubation or hatching temperature, which affects the development of bones, muscles, and tendons and results in legs that are not strong enough for the hatchling to stand on. Lack of strength may also result from an inadequate breeder-flock diet.

During or after the hatch, a common cause is a hatcher or brooder floor that's too smooth for hatchlings to walk on, so their legs slide out to the side. As a result, the leg muscles don't get sufficient exercise to develop properly; the birds

cannot walk; and, if the condition is not corrected, the birds will die. Paper toweling lining a smooth-bottom hatcher and covering the brooder floor for the first few days prevents legs from splaying.

Treatment. A bird with splayed legs must have its legs hobbled, or mechanically brought together, to help it stand until the muscles develop sufficiently to let it stand and walk on its own. The hobble should be made of something soft and flexible that will not cut into the leg or bind, as string or yarn would.

With goslings I have had great success using a small rubber band with a knot tied in the middle to create a figure eight, then slipping one loop onto each leg. After a day or two, most goslings can walk on their own, at which time the rubber bands must be removed so they won't cut into flesh as the goslings grow.

First-aid tape is another option. A Band-Aid is often handier, and the gauze pad may be used as a spacing guide to get the legs the appropriate distance apart. A smaller Band-Aid is needed for bantams than for larger birds.

In both cases cut the Band-Aid in half lengthwise to make it the right width to fit the bird's leg. To avoid getting the tape or Band-Aid too tight and cutting off circulation to the foot, protect the legs with a bit of double-sided foam mounting tape wrapped almost all the way around except for the inner sides of each leg. Then cover the foam with first-aid tape or a Band-Aid, making sure the sticky side is completely covered so it won't stick to down when the bird squats to rest. Keep an eye on the

Splayed leg may be caused by a too-high temperature or a slippery surface in the hatching tray or brooder.

color of the feet to make sure the hobble isn't cutting off circulation.

Depending on how long the bird has been sliding around with its legs out from under, it may have a hard time getting used to having its legs properly positioned underneath. If the hobble causes the bird to fall over, space the legs a little farther apart than they normally would be, and each day retape them a little closer together.

SPLAYED LEG REPAIR

Splayed leg may be strengthened with a hobble made from double-sided foam mounting tape wrapped with first-aid tape or a Band-Aid.

During recovery the bird must have a nonslip surface to walk on. Training a hobbled hatchling to walk may take several days, several reapplications of the hobble, and lots of patience. The sooner the hobble is applied, while bones and muscles are still flexible, the more quickly the hatchling will learn to walk properly.

Deformed Beak or Bill

Occasionally, a bird hatches with a deformed beak or develops a deformity as the bird grows. The upper and lower halves of the beak may grow in opposite directions, a condition called a crossed beak or scissors beak. Or the lower or, more typically, the upper half gets longer than the other half. In all cases the bird has trouble eating. The first thing to realize is that the condition is likely to be hereditary, so even though such a bird may live, any eggs it lays (if a female) or fertilizes (if a male) should never be used for hatching. Such birds often have other problems as well and may be picked on by flock mates. Sometimes the kindest thing to do is to put down such a bird.

Treatment. A beak or bill deformity may not be readily apparent at hatch but gets worse as the bird matures and its beak grows longer. Depending on the severity of the deformity, the bird may or may not be able to eat properly. If the bird is a chick, poult, or keet and you want to keep it as a pet, you may be able to mitigate the condition with periodic judicious trimming. The upper half is naturally a little longer than the lower half. If the upper half begins to overlap the lower, simply shorten it with a fingernail file. Once it has passed the filing stage, use fingernail or toenail clippers to trim it back using the same method you would use for trimming fingernails.

If you don't let the beak grow too far out of whack, the part that needs to be trimmed away will be lighter in color than the rest of the beak. Trim a little at a time to make sure you don't get into live tissue and cause pain and bleeding. When in doubt, look at the beak from the inside and you should be able see where the live tissue ends. Keep an eye on the beak's growth, and trim it as often as necessary for the bird's comfort. Chances are you'll

BEAK DEFORMITIES

crossed beak   upper beak too long   correct beak

have to trim the beak periodically throughout the bird's life.

Waterfowl sometimes have bill deformities as well, but their bills cannot be trimmed because they are edged with lamellae — tiny toothlike ridges used to bite off bits of vegetation, to grab and hang onto insects, and to strain food particles out of water. The kindest thing to do for such a duckling or gosling is to put it out of its misery.

Stargazing

Congenital loco, known as stargazing, is a nervous condition of chicks and poults that causes a hatchling's head to bend so far back that its beak points

Stargazing is a nervous condition that causes a chick's or poult's head to bend so far back its beak points skyward.

toward the sky. The epileptic-like muscle spasms that pull the head back tip the bird over. After a few days of flopping around, it dies from lack of food and water. Hatchlings exhibiting this condition are discarded by commercial hatcheries and dedicated breeders. Because they're rarely kept for study, no one is certain how common the condition is or what causes it. The few studies that have been done indicate it may result from a defect in the ear structure.

Some evidence indicates that congenital loco and stargazing may be two different things, inasmuch as congenital loco appears at the time of hatch and is lethal, whereas a similar stargazing condition may appear several days after hatch, last a few days, then go away. One of my New Hampshire chicks began stargazing when it was startled, for instance, by a loud noise or when I raised the heater panel to check underneath. The chick's head would snap back, its beak pointing straight in the air. Sometimes the head wobbled from side to side. Other times the chick moved in a backwards circle. The episodes were always brief and occurred over about a 10-day period, after which they stopped and the chick became so normal I could no longer identify it among others in the brooder.

SKEWED GENDER RATIO

Unlike in humans, with birds the female determines the gender of the offspring. Opposite to humans, male birds have two similar sex chromosomes (designated as ZZ), while females have dissimilar ones (ZW). Genes carried on this single pair of chromosomes determine gender, as well as any characteristics that are sex-linked.

Each egg inherits a Z chromosome from the male. Half the eggs inherit a Z from the hen, the other half inherit a W. Every fertilized egg contains a sex chromosome from the male, but a hen transmits her female sex chromosome to only about 50 percent of the eggs she lays. Since each egg has a 50/50 chance of containing two sex chromosomes, theoretically, the sex ratio should be approximately 50 percent males and 50 percent females. That would be the case if 100 percent of the eggs a hen lays are fertile and 100 percent of them hatch.

Deviations from 50/50

However, not all eggs are fertile (even when a male is present), and not all fertile eggs hatch. The two most common reasons for significant deviations from this ratio are sex-linked lethal genes and the random death of embryos and chicks.

If the unfertilized eggs and early-death embryos would have been males, then the hatch will have more females than males. That does sometimes happen. But more often, for reasons not fully understood by scientists and poultry keepers, the unfertilized eggs and early-death embryos would have been females, resulting in a hatch with a higher percentage of male hatchlings than female hatchlings.

The exact sex ratio of any particular group of hatchlings is influenced by many factors and varies with individual hens and the quality of the breeder male or males. Generally among high egg-producing strains, the ratio does lean toward 50/50. Among strains that are not bred for high egg production, the tendency leans toward more males — in some cases running as high as 70/30.

Both Sexes in One Bird

On rare occasions a bird will be a gynandromorph, meaning it has both male and female characteristics and organs. A dramatic example is the bilateral gynandromorph or half-sider; as the bird develops, one side of its body looks like just a male and the other side looks like a hen.

Not all gynandromorphs are so clearly distinguishable. I once had a mallard that grew up looking like a drake. No, a duck. No, a drake. I couldn't determine which it was. Sometimes it looked and acted male, sometimes female. I decided to remove it from my breeding flock by having it for dinner. In the process of cleaning the bird, I made an amazing discovery: it was both.

Issues Affecting Hatchability

Just because a fertile egg contains a sperm cell, making it potentially capable of development during incubation, doesn't mean the egg will survive the incubation period and hatch into a healthy bird. Even a hen averages a hatch of only 89 percent of her fertile eggs. For artificial incubation in a well-designed unit that is properly run, the average rate is 85 percent for land fowl, 70 to 75 percent for waterfowl.

By conventional definition anything more than 75 percent of fertiles hatched is considered a good rate; less than 50 percent is low. If you are hatching in the mid- to high range, you can probably improve your rate by fine-tuning the way you run your incubator. How seriously a hatch will be affected by irregularities in an incubator's operating variables — temperature, humidity, ventilation, turning, and egg orientation — depends on these factors:

- How considerable the irregularity is
- How many variables are involved
- The stage of embryonic development at the time
- How long the irregularity lasts

If your hatches consistently fall in the low range, look for other causes. Good incubator sanitation is important for hatching success, as well as for giving hatchlings a healthy start in life. Because hatching is a major source of incubator contamination, take time to clean your incubator thoroughly after each hatch. At the end of each hatching season, do an especially thorough job of cleaning your incubator. Cleaning and disinfecting will not destroy all disease-causing organisms but will make the environment less favorable for the survival of any microorganisms that remain. If you let cleanup go until the beginning of the next season, you'll have more microbes to contend with, harbored all that time in fluff and other debris.

Assuming your incubation techniques and sanitation procedures are up to snuff, the next thing to investigate is your breeders. To begin with, the eggs of better layers generally have a higher rate of hatchability than those of poorer layers. The small eggs with small yolks laid by young hens are also lower in hatchability and produce a high percentage of deformed embryos compared to eggs laid by mature hens. Hatchability rises as eggs reach full size but drops slightly between a hen's first and second year and continues a gradual decline as the hen ages, especially if the hen is out of condition, parasitized, or diseased.

BREEDER FLOCK HEALTH

A good indication of disease is the regular appearance of eggs with rough, misshapen, or thin shells with a low rate of hatchability and a high rate of dead embryos and dead hatchlings. Diseases may be transmitted from infected breeders to their offspring through hatching eggs in one of two ways: The infectious organism may (rarely) enter the egg as it is being formed in an infected hen, or it may get on the shell as the egg is laid or when it lands in a contaminated nest. Bacteria then enter the egg through the shell, which occurs more readily if the shell cracks or gets wet (for example, during improper cleaning). Diseases also spread from infected hatchlings to healthy ones in the incubator (often inhaled in fluff) or in the brooder (usually through ingested droppings in feed or water).

Salmonella bacteria are fairly common in poultry and can cause a low rate of hatchability and high rate of hatchling deaths. The most common diseases caused by *Salmonella* are pullorum and fowl typhoid, both of which may be eliminated from a breeder flock through blood testing to identify and cull carriers. Paratyphoid, of which many types exist, is another disease caused by *Salmonella*, but one that is more difficult to detect.

Additionally, eggs from breeders that have recovered from a disease may be infertile, because some diseases can cause permanent damage to the ovary. Especially troublesome to fertility are chronic respiratory disease, infectious coryza, infectious bronchitis, Marek's disease, and endemic (mild) Newcastle disease.

The accompanying table lists the most likely diseases to affect hatchability. It is included here, not to scare the bejeebers out of you but to make you aware of possible causes should you be unfortunate enough to experience large percentages of dead embryos or hatchlings. Fortunately, disease is unlikely to be an issue in the average backyard small flock. A more likely cause of low hatchability is poor nutrition.

Eggs from Outside Sources Are Risky

Incubating eggs collected from various sources is a sure way to introduce disease-causing organisms into your incubator. To avoid bringing diseases into your existing flock, hatch only eggs laid by your own birds. If you wish to hatch eggs from outside sources, incubate eggs from one source at a time, then thoroughly clean and sanitize your incubator before starting to incubate the next setting.

Diseases Affecting Hatchability and Hatchling Deaths

PROBLEM	DISEASE	CAUSE	SOLUTION
Eggs appear green when candled	Aspergillosis	*Aspergillus fumigatus* fungus	Do not set eggs with cracked or poor shells; clean and disinfect incubator between hatches
Embryos die on days 1–7	Aflatoxicosis	Toxins produced by molds in grain and bedding	Avoid feeding moldy grain to breeders; remove and replace moist or moldy breeder bedding
Embryos die on days 19–21; watery yellowish-brown yolk sacs	Colibacillosis (yolk sac infection)	*Escherichia coli* bacteria	Set only clean eggs
Embryos die on day 21 without pipping	Pseudomonas	*Pseudomonas aeruginosa* bacteria	Improve incubator sanitation
Many embryos die in shell, pipped or not	Paratyphoid	Several types of *Salmonella* bacteria	Collect hatching eggs often; set only clean eggs
Embryos die on day 21, pipped or not pipped; or hatchlings have distended abdomens, swollen navels, or bad odor and may die during first week	Omphalitis (mushy chick disease)	*E. coli* bacteria	Set only clean eggs
Hatchlings have white or greenish-brown pasty droppings; large percentage die up to four weeks of age	Pullorum	*Salmonella pullorum* bacteria	Blood-test breeders to eliminate carriers
Large percentage of hatchlings die suddenly	Fowl typhoid	*Salmonella gallinarum* bacteria	Blood-test breeders to eliminate carriers
Hatchlings are weak and droopy; large percentage die by ten days of age	Paratyphoid	*E. coli* bacteria	Clean and disinfect incubator and brooder between hatches
Gasping; swollen eyes; paralysis in hatchlings to three weeks of age; up to 10 percent die	Aspergillosis (brooder pneumonia)	*Aspergillus fumigatus* fungus	Do not set eggs with cracked or poor shells; clean and disinfect incubator and brooder between hatches

Adapted from *The Chicken Health Handbook*, 1994, by Gail Damerow

BREEDER FLOCK NUTRITION

Nutritional deficiency is one of the most common causes of poor hatchability. Lay ration contains too little protein, vitamins, and minerals to provide adequate nutrients in hatching eggs, to the detriment of embryo development. The older the breeders are, the more they need a higher level of nutrition than can be provided by lay ration alone.

An embryo's appearance, combined with the day on which it died, provides a clue regarding nutrients that might be lacking. In riboflavin deficiency, for example, deaths peak at three points: the fourth, tenth, and fourteenth day of incubation. Embryos may be dwarfed, have beaks that look like a parrot's, have unusually short wings and legs, and have clubbed down.

The accompanying table lists typical deficiencies, their signs, and the stage at which each appears in chicken eggs. Other species have similar manifestations, appearing at the same stage of development, percentagewise; for example, day 10 is about halfway through incubation for chickens, which would be day 14 for others. How likely these abnormalities are to occur, as well as the stage at which deaths peak, depends on how serious the deficiency is or how recently before egg collection the deficiency was corrected.

In some areas feed stores carry a breeder ration. The next closest thing is a game bird ration. Since feeding grain dilutes protein, vitamins, and minerals, eliminate grain from the breeder-flock diet. Also add a vitamin/mineral supplement to the drinking water. Vitamins A and E are particularly important for males; a deficiency in either vitamin reduces fertility.

Parrot beak is an infrequently appearing deformity that may be hereditary or may be caused by a nutritional deficiency in the breeder flock.

Switch your flock to a breeder diet at least one month before you plan to collect hatching eggs, and continue it throughout the breeding season. You should see at least a 10 percent improvement in the hatchability of eggs from your flock.

HEREDITARY DEFECTS

Each bird has a combination of dominant and recessive genes. When a dominant gene pairs up with a recessive gene, the dominant gene overshadows or modifies the recessive gene, and the dominant trait prevails. Hereditary defects are caused by recessive genes that become apparent when two birds that carry the same recessive gene are mated.

Revealing recessive traits can be a good thing or a bad thing. If the recessive is desirable, you want to encourage it. If it is undesirable, you want to weed it out, which is possible only by maintaining sufficient genetic diversity to prevent the concentration of undesirable recessives in all your breeders.

Not all traits are controlled by dominant or recessive genes but rather by combinations of genes. An example is rumplessness, a genetically complex feature that is typical of Araucanas but occasionally appears in other chicken breeds. Rumplessness is determined by an interaction among many different genes. The more inbred your flock becomes, the more likely it is to reveal recessive traits.

Inbreeding

Continuous, close inbreeding causes a phenomenon known as inbreeding depression, for which low hatchability is usually the first sign. Later signs are fewer and fewer eggs laid and chicks lacking in constitutional vigor, meaning they're droopy and unthrifty, and they may or may not die soon after hatching.

Inbreeding is unavoidable if you breed birds for exhibition or you're trying to preserve one of the rarer classic breeds. By concentrating genes

Hatching Problems Related to Breeder Nutrition

SIGN	DEFICIENCY
DAY OF DEATH	
Early incubation	Vitamin A
1st–7th day	Biotin
4th day	Riboflavin, vitamin E
10th day	Riboflavin
10th day–21st day	Magnesium
14th day	Riboflavin
17th day	Vitamin B_{12}
18th–19th day	Vitamin D
19th–21st day	Biotin
20th–21st day	Manganese
Late incubation	Folic acid, pantothenic acid, vitamin E, vitamin K, selenium excess
At pipping	Folic acid, selenium excess
Soon after hatch	Vitamin E
APPEARANCE OF EMBRYO	
Beak/head abnormal	Zinc
Beak short	Vitamin B_{12}
Upper or lower beak short	Vitamin D
Crooked (parrot beak)	Riboflavin, manganese
Bones or beak soft and rubbery	Vitamin D
Clubbed down	Riboflavin
Eyes pale	Vitamins A and D
Small	Zinc
Feathers abnormal	Pantothenic acid
Fluid in body	Vitamin B_{12}
Growth dwarfed	Riboflavin
Stunted	Vitamin D
Legs/feet/wings twisted	Biotin
Legs bowed	Vitamins A and D
Legs short	Riboflavin, manganese
Legs undeveloped	Vitamin B_{12}
Legs missing	Zinc
Slipped tendon	Vitamin B_{12}, manganese
Navel not closed	Iodine
Skull deformed	Biotin, manganese
Spine poorly developed	Zinc
Incubation time too long	Iodine

Source: *The Chicken Health Handbook*, by Gail Damerow, Storey Publications, 1994.

inbreeding creates uniformity of size, color, and type but also brings out weaknesses, such as reduced rate of lay, low fertility, poor hatchability, and slow growth.

For every 10 percent increase in inbreeding, you can expect a 2.6 percent reduction in hatchability. A flock that reaches 30 percent hatchability is nearing extinction due to failure to reproduce. A flock can become extinct after only six to eight generations of brother-sister matings. You can slow inbreeding depression and improve reproduction in the following ways:

- Select your breeders in favor of number of eggs laid, hatchability of the eggs, and vigor of the resulting offspring
- Avoid brother-sister and offspring-parent matings and instead mate birds to their grandsires or granddams
- Keep no fewer than 50 birds in your breeder flock, allowing for more distant matings and more gradual inbreeding
- Retain as future breeders those with the best fertility, hatchability, chick viability, disease resistance, and body size
- Do not breed birds that have any tendency toward infertility
- Occasionally introduce birds from a different strain into your breeder flock

Some strains are less susceptible than others to the effects of inbreeding depression. And popular breeds with lots of varieties offer more opportunities for avoiding inbreeding depression than less popular breeds with fewer numbers or breeds with few varieties. Mating birds from different strains invariably results in hybrid vigor. The opposite of inbreeding depression, hybrid vigor causes more rapid growth, larger size, improved productivity, and increased viability.

Lethal Genes

One of the results of inbreeding is the concentration of lethal genes, which can cause a bird to die, typically as an embryo during incubation. In chickens alone, more than 50 different lethal genes have been identified, most of which are recessive. When two birds are mated that carry the same lethal recessive, 25 percent of their offspring will die. Embryos that die due to a lethal gene, as well as those that survive until hatch, usually have some sort of obvious deformity.

The creeper gene — the most widely studied lethal gene — is carried by short-legged Japanese bantams. When a creeper hen is mated to a cock carrying the creeper gene, one-quarter of the chicks die during the first week of incubation.

Dark Cornish carry a similar short-leg gene that causes embryos to die during the last days of incubation or to pip but be unable to break out of the shell. Signs of Cornish lethal include short beaks and wings and bulging eyes.

Congenital tremor is a lethal gene found in a number of breeds, including Ancona, Plymouth Rock, Rhode Island Red, White Leghorn, and White Wyandotte. Chicks hatch but can't control their neck muscles. When a bird tries to stand, its head falls over and the chick falls down. Unable to eat or drink, it dies soon after hatching.

Crestedness in ducks is a mutation in which the crest grows from tissue coming through a hole in the bird's skull. When a crested duck is mated to a crested drake, 25 percent of the offspring develop a brain outside the skull and fail to hatch.

A good way to find out about lethals and other problem genes in your chosen breed is to chat with experienced breeders, join a breed club, look for books about your breed, and do an online search using as key words your breed's name plus the term "lethal gene." Fortunately, as numerous as they are, lethal genes are not all that common.

11

HATCHLING IDENTIFICATION

When you hatch eggs from different hens or different matings, you may want to keep track of the parentage of each hatchling. For example, you may be involved in a conservation project — working to build the population of a rare breed and thus helping save the breed from extinction, and need to track matings so you can preserve the greatest genetic diversity for your breed. Or you may be trying to identify the matings that produce show-quality birds.

By tracking, or pedigreeing, your matings with certainty, you'll be able to identify problem breeders, such as a cock that is not optimally fertile or a hen that typically produces weak chicks. And you'll be in a better position to control matings to improve the genetic strength of your flock's future generations.

Embryo Dyeing

One way to keep track of hatchlings is to dye the embryos different hues so they are color coded when they hatch. Embryo dyeing works best on birds with white or light-colored down. For darker breeds a more suitable method is to separate the eggs before they hatch, as described in Egg Separators on page 207.

Embryo dyeing involves using a hypodermic needle to inject dye into incubating eggs, allowing you to color-identify chicks from different matings. Dyeing in no way affects a hatchling's health or growth rate, if you handle the eggs carefully and use only clean materials.

To dye embryos you will need a 22-gauge ¾-inch (19 mm)-long hypodermic needle (commonly used for dogs and cats; you can probably get one from your vet), a sharp sewing needle of the same size or a little bit bigger (or a 20-gauge hypodermic needle), and a set of food dyes in 2 or 3 percent concentration. Red, green, blue, and purple (made by combining red and blue) show up best. Yellow and orange don't alter down color enough to work well. To avoid inadvertently mixing two colors, use a clean needle and syringe when you change to a different color dye.

DYEING AN EMBRYO

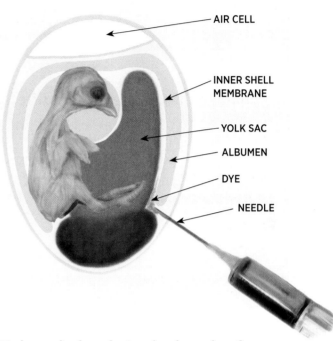

AIR CELL

INNER SHELL MEMBRANE

YOLK SAC

ALBUMEN

DYE

NEEDLE

Embryos dye best during the eleventh to fourteenth day of incubation. To avoid chilling the eggs, remove no more eggs from the incubator than you can dye within 30 minutes.

Talking Fowl

leg band. A method of identification consisting of a plastic or aluminum ring that wraps around a bird's leg

pedigree. A complete record of a bird's ancestry

pedigree basket. A basket-like enclosure used in an incubator or hatcher to separate chicks resulting from different matings

toe punch. A device used to punch holes into the webbing of a hatchling's feet for purposes of identification

wing band. A method of identification consisting of an aluminum strip that is attached through a bird's wing web, commonly at the time of hatch

STEPS FOR TINTING EMBRYOS

1. Approximately ½ inch (13 mm) from the tip of the pointed end of the egg, disinfect an area about the size of a quarter by wiping it with 95 percent rubbing alcohol or povidone-iodine (such as Betadine).

2. Dip the sewing needle into the alcohol or iodine.

3. Cushioning the egg in one hand, make a tiny hole in the center of the disinfected area by pressing against it with the needle, twisting the needle back and forth until it just penetrates the shell and membranes. Take care to make only a tiny hole that does not go deeper than necessary to pierce the inside membrane (no more than ⅛ inch [3 mm]) or the embryo may die.

4. Dip the hypodermic needle into the alcohol or iodine and fill it with ½ cc of dye.

5. Insert it into the hole so the tip is just beneath the inner shell membrane. Slowly depress the plunger to release the dye without letting it overflow.

6. Seal the hole with a drop of melted wax or a small piece of Band-Aid — a sheer strip sticks best.

7. Return the eggs to the incubator, and continue incubating them as usual.

The hatchlings will appear in a rainbow of colors, each having down of the colors you gave it. The dye will start wearing off in about 2 weeks, as the hatchlings grow their first feathers. At that time you'll need to apply some other form of identification, as described in Identification Methods, page 209.

Step 3

Step 5

Step 6

Within a few weeks dyed down is replaced by feathers of a normal color for the breed, by which time some other form of identification must be used.

Egg Separators

You can keep track of which bird hatches out of which egg in your incubator by separating the eggs before they hatch so they won't get mixed together. One option is to use hardware cloth to divide the hatching tray into compartments. This system works if each compartment has its own cover. If you rely on a cover that fits over the entire hatching tray, when you lift it off to retrieve hatchlings, the more rambunctious among them may scramble over the dividers and get mixed up.

On several occasions when I've needed to keep track of two different groups of hatchlings in my cabinet incubator, I put one group of eggs in the covered hatching tray at the bottom and the other group in the top setting tray. I used one of the empty setting trays as a cover for the top tray and placed paper towels in the other empty setting tray so hatching debris from the top tray wouldn't fall onto the eggs in the bottom tray. This system works, of course, only when the automatic turner is turned off during the hatch.

USING PEDIGREE BASKETS

The standard method of separating eggs during a hatch is to use pedigree baskets, so called because they allow you to track the pedigree of each individual egg. I have never seen them available for sale anywhere, which likely has to do with the great variety of incubator sizes and style. Baskets must fit the hatching tray in your specific incubator. Another reason has to do with the infinite number of combinations in size and number of eggs you might wish to keep track of. As long as the baskets fit your hatcher, they may be fashioned for any number of eggs of whatever size your flock produces.

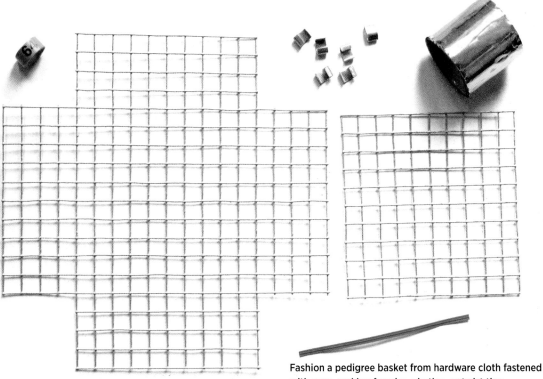

Fashion a pedigree basket from hardware cloth fastened with cage-making ferrules, zip ties, or twist ties.

Ready-made items that might be used as pedigree baskets include small plastic or coated wire baskets set upside down over groups of eggs. Small plastic or wire baskets might be found at an office supply store, a school supply store, a toy store, a drugstore, or a dollar-type store. (For my take on pedigree hatching bags, refer to the Screwpot Notions section in the appendix.) Before you go shopping, measure the height of your hatcher so you won't get baskets too tall to fit or that will interfere with any of your incubator's operating mechanisms.

To keep rambunctious hatchlings from tipping the upside-down baskets and slipping out, fasten the baskets to the hatching tray with a couple of twist ties — just be sure to untie one basket at a time and remove its occupants from the hatcher before untying the next basket. Alternatively, you might fashion lids for the baskets using something with an open weave, such as window screening or hardware cloth, that won't impede the hatcher's airflow.

We make our baskets from ½-inch (13 mm) hardware cloth, bent to form boxes that are 2 inches (50 mm) high. We chose a height of 2 inches because it fits nicely in the covered hatching tray of our incubator. Each basket holds 1 week's worth of eggs from a single hen. The measurements are determined by placing that number of eggs in a square on a flat surface — adding two or three extras to give hatchlings a little wiggle room — and measuring how much space the eggs take up.

To form corners for the basket, bend the hardware cloth on the edge of a table or workbench.

A hardware cloth cover is cut to fit and hinged to the basket on one side. We use cage-making ferrules to fasten the sides together and to hinge the lid — one ferrule at each corner and two to hinge the lid, but cable ties (zip ties) work just as well. Twist ties work, too, but don't last as long so must be periodically replaced. But a twist tie makes a handy and easily removed clasp for fastening down the lid after eggs have been placed in the basket.

To prevent a chick or yourself from getting scratched by the cut-off wire ends, fold tape all the way around the edges of the basket and the lid. We use aluminum foil tape because it's designed to withstand moisture and therefore won't peel off in the incubator's high humidity.

Chicks hatching in a pedigree basket remain separated from other chicks scrambling around on the hatching tray.

TRACKING PEDIGREES

If you have two groups of eggs to keep track of, you need only one basket to separate one group from the others in the hatcher. If you use two or more identical baskets, you'll need to identify each one so you can keep track of which basket holds which eggs. A numbered band fastened to the top

Pedigree Record

Pedigreeing makes sense only if you keep a detailed record for each hatchling. Your records should include the following minimum information for each bird:

- Breed
- Dam (hen that laid the egg)
- Sire (male that fertilized the egg)
- Date of hatch
- Sex (as soon as known)
- Color
- Identification number or leg-band color code

of each basket offers a handy way to keep track of which is which. Alternatively, baskets fastened together with zip ties might be color coded by the color of the ties used for each, or spiral leg bands may be used for color coding.

As you collect eggs from different matings, use a grease pencil or china marker to identify them according to whatever system you devise to identify your various matings. You might, for example, assign specific letters or numbers to each mating combination and create a master list to help you keep track of what each set of letters or numbers represents.

On the eighteenth day of incubation, when you would normally stop turning the eggs, group them in your baskets, making a note of which mating is represented by each basket. After the chicks hatch, remove one basket at a time and use some method of identification, as described below, to keep track of which chicks came out of which basket.

Identification Methods

As the hatchlings are removed from the incubator, they need some form of identification by which you can tell them apart. Temporarily, they may be marked with a dab of food dye, nontoxic marker, or children's finger paint to the head, back, or rump.

Gel or liquid food dye works better than the paste form in a tube, which in my experience doesn't stick well to start with and rubs off too quickly. Nontoxic markers and finger paints may be found anywhere school supplies are sold. All these options work best on light-colored landfowl, but avoid using red, which looks like blood and can invite picking. Marks on ducklings or goslings don't last long, especially once they start bathing, but will last at least long enough for you to apply a more permanent method of identification soon after the hatch.

For darker breeds different colors of nail polish applied to the feet make an alternative form of temporary identification. Like other marks, the polish wears off rather quickly, which is a good thing because it can invite toe picking. Because red shades attract picking, they are not the best choice for this purpose.

You may, of course, apply a more permanent method of identification as soon as you remove the birds from the incubator. Permanent methods include leg banding, wing banding, and toe punching.

LEG BANDING

Banding the legs of hatchlings is tricky business, both because their legs are so delicate and because as the hatchlings grow they quickly outgrow their leg bands. Leg banding therefore requires changing the bands to a larger size as often as necessary to keep them from binding.

A leg band consists of a plastic or aluminum ring that wraps around a bird's leg. Plastic bands are more suitable for the tiny hatchling legs; aluminum bands may be applied once they are full grown. Plastic bands come in different sizes and colors, with numbers (called bands or bandettes) or without (called spiral bands or **spirals**). The latter are used to identify matings by color code, and since they are thinner than bandettes, they are easier to apply on little legs. You might start with spirals and change to bandettes later as the birds, and their legs, get bigger.

Leg-band sizes range from #2 to #16. The size denotes the band's inside diameter in sixteenths of an inch. The American Bantam Association uses its own sizing system in millimeters.

A leg band must be the right size for the bird to ensure it is neither so small it binds nor so big it falls off. Different species and breeds require different sizes. Within a breed the males usually require a larger size than the hens, and younger birds need a smaller size than mature birds. For the large breeds, start with #4 bands; for bantams and keets, you'll need #3 or maybe #2 for the really little guys.

Since bandettes and spirals break, especially if they're reused several times, a good idea is to apply more than one, in case one gets lost. You might put

Zip ties (cable ties) make handy color-coded leg bands, but must be checked often for fit and replaced as the birds grow.

a bandette of the same color and number on both legs or, when using color-coded combinations, put the same combination of spirals on both legs.

Nylon zip ties (also called cable ties or tie wraps) come in many different colors and are easier to apply to the legs of hatchlings. Unlike leg bands, they won't break and get lost but must be cut off and replaced as the birds grow. Be sure to leave the ties loose enough to get scissors in to cut them off for replacement, but not so loose a bird could get its beak or bill or other foot under the tie. And trim off the excess to keep birds from tightening a tie by picking on the end.

Bands and zip ties both require several size changes by the time birds are full grown. Replace bands or ties as they become too tight to slide up and down easily. Never leave a band or tie on so long that it binds. If either is left too long on a growing bird, the leg will grow around it and the band will become embedded in the leg, resulting in lameness. Similarly, as a cock's spurs begin to grow, make sure the leg band remains above the spur.

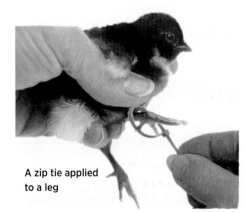

A zip tie applied to a leg

Cutting off the excess

WING BANDING IS EASIEST

Wing banding is a one-shot deal, and the band remains on the bird for life, unless you deliberately remove it. Wing bands come in colors, as well as plain aluminum, and are usually stamped with sequential numbers or embossed to your specification. The band is applied through the wing web, a flat triangular flap of skin that becomes evident when the wing is fully extended. To avoid tearing the tender wing web, apply wing bands after hatchlings have completely fluffed out and have begun to toughen, usually at one day old.

Some bands are designed to be applied by hand, pinched shut between your fingers. Other more tamperproof styles require a plierslike applicator to snap them closed by means of a rivet or eyelet. Both styles are easy to apply, if you are careful not to tear the edge of the wing web. Just make sure the band doesn't slip over the end of the wing and restrict wing development as the bird grows. Continue checking bands until the wings get big enough for the bands to stay in place. A detailed video on how to apply a wing

Make sure the band remains properly in place (right) and doesn't slip over the end of the wing (left), which would bind as the wing grows.

band may be found online at http://www.national-band.com/WingBandingDirections.mov.

Except in tightly feathered breeds, wing bands are clearly visible only on close examination. They may be used for any species or breed, although show judges often frown on a little bantam wearing a visible wing band.

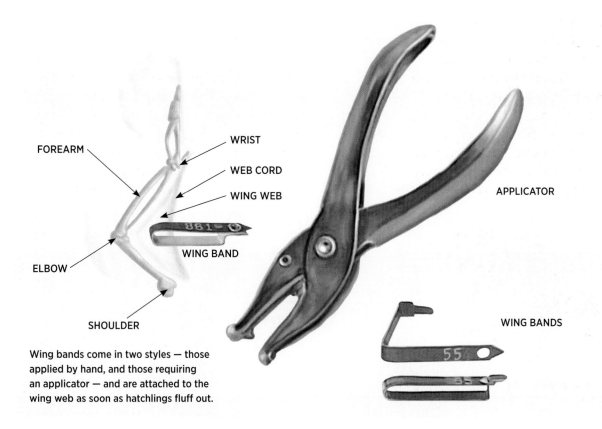

FOREARM
WRIST
WEB CORD
WING WEB
WING BAND
ELBOW
SHOULDER

APPLICATOR

WING BANDS

Wing bands come in two styles — those applied by hand, and those requiring an applicator — and are attached to the wing web as soon as hatchlings fluff out.

TOE PUNCH

Toe punching involves punching holes in the webs between the toes in a coded pattern. A toe punch, available from many poultry supply catalogs, is about the size of fingernail clippers and functions like a paper hole punch. It is used to make a hole through the web. Please note that, although it's called "toe" punching, you don't punch the toes themselves but the web between the toes.

How It Works

Each foot has three main toes and therefore two webs: the outer web (between the middle and outside toe) and the inner web (between the middle and inside toe). The patterns formed by punching or not punching a web form 16 possible combinations that let you identify up to 16 different matings. Each pattern is assigned a number from 1 to 16, and all hatchlings from one mating are punched with the same pattern.

With the chick facing the same direction you are, and starting at the left side, the first web (left outer) stands for 1; the next web (left inner) stands for 2; the next web (right inner) stands for 4; the far right-hand web (right outer) stands for 8. Assign each mating a number from 1 to 15 and identify hatchlings from each mating by adding up the numbers corresponding to the punched webs.

Thus hatchlings from mating number one have the left outer web punched. Those from mating number two have the left inner web punched. Those from mating number three have both the left outer and left inner webs punched (1 + 2 = 3). Those from mating number 5 have the left outer and right inner webs punched (1 + 4 = 5). And so forth. A sixteenth mating may be identified by leaving all the webs unpunched, but this option is seldom used, as you never know for sure if intact webs grew back or were never punched.

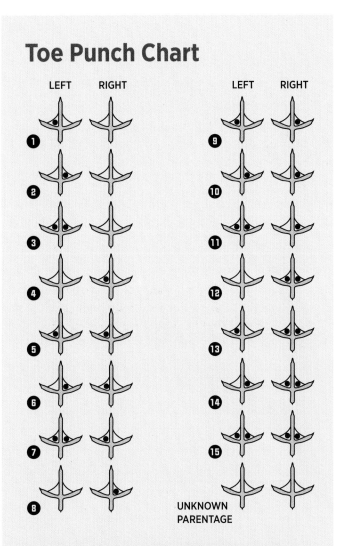

Toe punching puts a number-coded hole through the web between toes, where it is readily visible only on close inspection.

A toe punch is about the size of fingernail clippers and functions like a paper hole punch.

Punch webs as soon as the hatchlings are removed from the incubator. Hold each one gently but securely in one hand with the chick facing the same direction as you and one foot extended at a time. Carefully center the punch over the web, and with one firm stroke, punch a clean hole through the web. If the punched-out bit of skin hangs up, it must be pulled off to keep the web from growing back together.

Although toe punching is widely done, it is not an ideal identification method for several reasons: the holes are not readily visible, except on close inspection, and are not at all visible on chickens with feathered feet; a web can grow back; when the birds mature and wander outdoors, a twig can get caught in a hole, tearing the web. For these reasons I don't care much for toe punching and no longer do it.

FUN AND EXCITEMENT BEGIN

Once you've chosen one of the several systems for keeping track of the pedigrees of your birds, and you have selected the organizational method that best suits your needs, you'll be in fine shape to start collecting eggs for hatching. That's when the fun and excitement begins. Soon you'll have a new growing flock of your own choosing and will be ready to start the hatching and brooding cycle all over again.

SCREWPOT NOTIONS

The notions presented below are commonly called old wives' tales, a term I don't care much for because they are just as often repeated by old husbands, as well as by young wives, young husbands, and people who aren't married at all. In my family these notions would be called screwpot ideas — a word that combines screwball and crackpot into one all-encompassing term. Some of these ideas do contain a kernel of truth, which seems to lend them a modicum of credibility. Most of them, however, either have never been scientifically proven to be true or have been demonstrated to be decidedly false.

The Internet is a great place to find accurate information about hatching and brooding poultry.

Yes. And no. It depends on where on the Internet you go. Forums that are frequented primarily by first-time poultry keepers tend to circulate a lot of misconceptions, but I've also found serious errors on a number of university and Extension websites.

Before taking advice found on any website and applying it to your incubation technique or your brooder management, verify the information against two or three other websites you consider to be reliable. If the wording is identical on every website you consult, keep looking — once something is posted online, other sites will copy it, whether it's accurate or not. So check it out before taking a chance on jeopardizing your hatch or your hatchlings.

You can prevent or cure coccidiosis by feeding (pick one) herbs, garlic, vinegar, milk, or yogurt.

No. No. No. No. And no. These notions are dangerous and can jeopardize the health, and lives, of your hatchlings. Certain herbs, as well as vinegar and yogurt, may be used in moderation to enhance the immunity and overall health of poultry. But too much garlic can cause anemia and therefore is hazardous to their health. Milk, in more than minor quantities, causes diarrhea, which is not a healthful condition in baby birds, especially if they are already sick.

Before modern drugs became available, milk sometimes was used as a flush to induce diarrhea to clean out the intestines of poultry infected with coccidiosis. Today, instead of spreading coccidial protozoa from the loosened bowels of infected birds, we have a variety of drugs called coccidiocides that destroy the parasites and reliably cure birds of the disease. A young bird with coccidiosis is a seriously ill bird. Don't gamble with its life by treating it with ineffective alternative "natural" products.

The only ways to prevent coccidiosis in the first place are through good sanitation and through vaccination or the use of coccidiostats, as described in chapter 5 on page 95.

The hatchability of shipped eggs will be compromised by security scan radiation.

I can find no study done specifically on hatching eggs passed through security scanners. I did, however, read one hatchability study in which eggs were subjected to high-intensity radiation for several minutes longer than an egg carton takes to go through a security scanner. One-third of the eggs failed to hatch, one-third of those that hatched soon died, and one-third hatched and survived.

People who routinely ship fertile eggs or hand carry them through security checkpoints report results that vary from no eggs hatch, to eggs hatch but the chicks are deformed, to eggs hatch without a problem — all of which leads me to believe the low hatchability of shipped eggs relates to other factors, as described on page 148. Suggestions that shipped eggs should be wrapped in foil to avoid radiation can backfire. A package that is subjected to inspection and found to be suspicious most likely will be opened, potentially delaying delivery.

A chick's gender may be determined by the way it acts or reacts.

Since every hatchling is either male or female, this concept is like taking a true-false test — you have a 50 percent chance of being right every time. Here are some of the tests that supposedly allow you to determine if a newly hatched chicken is male or female:

- **Hold out the chick by the loose skin on the back of its neck.** If it folds up its legs and goes limp it's a pullet; if it hangs loose or struggles, it's a cockerel
- **Hold the chick upside down by its legs.** If it goes limp, it's a cockerel; if it struggles, it's a pullet
- **Lay the chick on its back.** If it keeps kicking, it's a cockerel; if it stops kicking, it's a pullet
- **Drop a small, soft object, such as a square of toilet paper, over the chicks' heads.** Those that look up and chirp are cockerels; those that flatten down and keep silent are pullets.
- **Put down the feeder and drinker.** The first ones to find the feeder and drinker are pullets.
- **Keep a close eye on roosting behavior.** The first ones to roost are cockerels (or pullets, depending on who's telling the story).

The sex of an egg or hatchling can be determined by dowsing.

Dowsing is a technique for finding something that is not readily visible by observing how a pendulum swings in response to unseen influences. In this case, it is used to determine the gender of a fertile egg or hatchling. Over the Internet you can waste your hard-earned money on pendulum sex indicators and magnetic hatching egg sexers, some of which claim to come with a money-back guarantee. Old-timers used a needle and thread, a nail tied to a string, or a key hanging from a chain.

To perform this technique, you're told to dangle the needle, nail, key, or pendulum over a fertile or incubated egg (especially a pipped egg) or over the head of a hatchling. If the needle or nail swings back and forth, it's a male; if it swings in a circle, it's a female. Some people swear it works 100 percent of the time; others insist it doesn't work at all. Maybe

it depends on who's handling the pendulum. Along with using rods to dowse for water — which works for some people and not for others — let's file this one under unsolved mysteries.

An egg's shape determines the gender of the bird that will hatch from it.

If this gem were true, the poultry industry wouldn't spend untold hours (not to mention dollars) looking for easy ways to determine the gender of hatchlings so they can raise the males and females separately, which is an important aspect of industrial layer, broiler, and turkey production.

Most eggs have a rounded or blunt end and a more pointed end, although some eggs are nearly round, while others are more elongated. An egg's shape is established in the hen's shell gland, where the yolk and white are wrapped in shell membranes and then encased in a hard shell that takes on the shape of the shell gland itself. Each hen therefore lays eggs of a characteristic shape, so you can usually identify which hen laid a particular egg by its shape.

If an egg's shape determined the hatchling's gender, some hens would lay eggs that always hatch males and others would lay eggs that always hatch females. That, in fact, is a possibility, just as some human males produce all boys and some produce all girls. With birds, the hen, not the male, determines the offspring's gender, so it is possible that a particular hen might produce all males or all females, which the hen's owner would then accept as proof that an egg's shape determines the hatchling's gender.

Indeed, as far back as 350 BC, Aristotle (in *The History of Animals*) said: "Long and pointed eggs are female; those that are round, or more rounded at the narrow end, are male." But wait; in about AD 77 to 79, Pliny the Elder (in *Natural History*) said: "Eggs of a rounder formation produce a hen chicken and the rest a cock." Today's believers in this screwpot notion side with Pliny, despite the fact that in the early 1900s USDA researchers in Beltsville, Maryland, proved that the shape of an egg has nothing to do with the gender of the bird that hatches from it.

An egg's size or weight determines the gender of the hatchling.

False. In the early 1900s USDA researchers ran experiments to determine if large chicken eggs produce cockerels, while small eggs produce pullets. They found no significant difference between the weights of cockerels and pullets at hatch, and no relationship between chick weight as a percentage of egg weight, proving definitively that gender bears no relationship to egg weight.

The sex of a hatchling may be influenced by incubation temperature.

Maybe. According to this theory, a hen that broods during winter will produce more pullets than a hen that broods during summer. Likewise, eggs incubated at a temperature that is slightly lower than the manufacturer recommends will produce a greater percentage of pullets, while eggs incubated at a slightly higher temperature will produce a greater percentage of cockerels. The temperature deviation for this purpose is commonly quoted as being one-half degree Fahrenheit higher or lower than is considered optimum.

The question then comes up as to whether or not embryos switch gender. The usual explanation is that for some reason a lower temperature causes fewer male embryos to survive, while a high temperature causes fewer females to survive. But wait; in 1997 biologist Mark Ferguson of Britain's University of Manchester patented a precise technique for dropping the temperature of eggs for 3 days during early incubation, which causes 10 percent of the male embryos to change into females that mature into fully functional laying hens. However, when these hens are mated to normal males, the resulting fertile eggs all hatch cockerels.

You don't need to adjust an incubator's humidity levels.

This notion has two schools of thought: one is to keep the moisture level the same all the way through incubation and the hatch; the other is that you don't need to add any moisture at all. The latter is sometimes called dry hatching.

In an extremely humid climate, ambient humidity may be sufficient for incubation, and in fact increasing humidity within the incubator could be detrimental. But in an extremely dry climate, dry hatching would end in disaster.

How much tweaking an incubator's humidity requires depends not only on ambient humidity, but also on how many eggs the incubator (or hatcher) contains compared to the unit's full capacity. The eggs themselves give off a certain amount of moisture, so an incubator that's filled to its maximum capacity may require less humidity adjustment than an incubator that's only partially full. Bottom line: dry hatching or maintaining a constant humidity throughout may work in some, but not all, situations.

You don't need to turn eggs during the incubation period.

When you think about how eggs settle in a nest, how still a hen sits on them, and how infrequently she leaves the nest, you have to wonder who came up with the idea that eggs need to be turned at least three times a day (and automatic incubators turn them every hour). While working on this book I got curious about that, so I ran a couple of tests by incubating two groups of eggs in the same incubator at the same time. Half of them turned as usual and the other half not turned at all.

In both tests some eggs hatched whether they were turned or not. But both tests showed conclusively that turning increases the hatch rate by 70 to 80 percent. An analysis of the unhatched eggs revealed that about 6 times more embryos in the not turned group died during the first week or so of incubation, which backs the assertion that turning is most essential during early incubation. Interestingly, a setting hen leaves the nest more often during early incubation and presumably reorganizes the eggs when she settles back in on her return.

The unhatched eggs in the not-turned groups had a high rate of malpositioned air cells, resulting in a lot of dead-in-shells. And the shells of the eggs that did hatch were broken into smaller pieces than normally occurs, indicating that the chicks put up quite a struggle to get out.

So, yes, eggs may hatch even if they are not turned, but chances are greater that they won't hatch. At any rate, I've wasted enough eggs checking out this screwpot notion.

In event of a power outage, use an inverter plugged into your car's cigarette lighter to power your incubator.

An inverter is an electrical device that converts direct current (DC, such as is produced by a vehicle's battery) into alternating current (AC, the type of electricity that comes out of the outlets in your house). One of these little gizmos lets you use your vehicle's cigarette lighter to produce 120-volt AC household current and keep your incubator running.

Does it work? Yes. Is it a good idea? Not really. First of all, you have to make sure your inverter matches both your car's battery output and your incubator's power usage. Just how long you can power your incubator this way depends on the amps/hours of your battery and the wattage of your incubator. Second, you'll draw down your vehicle's battery unless you keep the vehicle running (in which case, you'll use up gas or diesel). Third, if you deeply discharge your battery often enough, you'll shorten the battery's life. And finally, cold weather reduces your battery's capacity, and power outages typically occur in wintertime.

Okay, so let's assume you're going to be driving your vehicle anyway, which helps keep the battery charged while it's powering your incubator. Bouncing around on the seat won't do the embryos much good, especially if you live in a rural area where potholes abound. And if the sun shines into the cab, especially if it hits the incubator, the temperature will soar and cook your chicks before they have a chance to hatch.

You could set up a stationary battery pack with an inverter and a battery recharger, but a small generator is a lot less trouble to set up and operate. I use one to power my incubator, as well as several brooders if they happen also to be in use at the time the power goes out. Whether your backup is a battery pack or generator, for safety's sake, arrange it in a covered area outdoors and power your incubator (and brooders) by means of a heavy extension cord.

Heating recycled egg cartons in a microwave or oven is a great way to sterilize them.

Oh, my. Just the thought of the possibility of introducing bacteria into my fridge gives me the heebie-jeebies, let alone running the risk of doing the same to my microwave or oven in the event the temperature and/or time factor should be inadequate. Nevertheless, lots of people sanitize cardboard egg cartons by putting them into a microwave. Somewhere between 30 and 60 seconds should do the trick, although this also carries the risk of unintentionally starting a fire. Less chancy is preheating the oven to 250°F (121°C) and heating cardboard cartons for 10 to 15 minutes.

Foam cartons are best sent to the recycle bin. They warp or melt too easily to be sanitized by heat, and they break apart too readily to be easily washed.

As you might guess, I'm not a fan of reusing egg cartons. Unless, of course, they are designed to be washable, such as those hard plastic egg boxes used by backpackers.

Plastic or fabric mesh is suitable for creating pedigree hatching bags.

Mesh produce bags, wedding veil mesh, mosquito netting and the like have long been suggested as inexpensive alternatives to pedigree baskets. You put the eggs into a mesh bag, or lay eggs on a circle of netting or mesh, gather it up, and tie it off with a twist-tie.

I have to admit I have never been tempted to use pedigree bags. I worry that a leg, wing, beak, or toenail might get tangled in the mesh. On rare occasion a hatchling has caught a hock, toe, or wing in the hatching tray of my incubator, where I found it struggling to get free. I can just picture what the outcome might be if such a baby bird got tangled up in a flexible mesh bag. No thanks. I'll stick with hardware cloth pedigree baskets.

GLOSSARY

ALBUMEN. Egg white

ALLANTOIS. The membrane arising from an embryo's gut that grows until it completely surrounds the embryo. Its function is to provide the embryo with oxygen, expel carbon dioxide, deliver nutrients from the albumen and calcium from the shell, and collect body wastes.

ALL-IN ALL-OUT INCUBATION. Single-stage incubation

AMNION. The innermost membrane that begins developing on the third day of incubation and surrounds only the embryo. It is filled with colorless amniotic fluid in which the embryo floats, giving it freedom to move and exercise, protecting it from mechanical shock, and preventing it from drying out.

ANTIOXIDANT. Any substance, such as vitamin A, that is capable of counteracting the harmful effects of oxygen, such as the damage caused by free radicals

APTERIA. Bare areas of a bird's skin between the areas on which feathers grow

AREA BROODER. A small area separated from a larger area for the purpose of confining chicks close to heat, feed, and water

ARTIFICIAL INSEMINATION. The introduction of semen into a hen other than by natural mating

ASEXUAL REPRODUCTION. The phenomenon whereby an unfertilized egg develops an embryo. Also called *parthenogenesis*

AS-HATCHED CHICKS. Chicks that have not been sorted by gender but are kept in the gender ratio in which they naturally hatched from a single setting of eggs. Typically that ratio is roughly 50 percent female and 50 percent male. Also called *straight run* and *unsexed*

AUTOSEX. A straightbred variety or breed that displays clearly distinct sex-linked color characteristics by which males may be easily distinguished from females at the time of hatch. The term autosexing was developed to differentiate sex-link purebreds from sex-link crossbreds

BANDETTE. A numbered plastic spiral applied to a bird's leg for identification purposes

BANTAM. A small breed, generally about one-fourth to one-fifth as heavy as its large-size counterpart

BARRED. A color pattern of individual feathers consisting of crosswise stripes of two distinct alternating colors

BATTERY BROODER. A series of box brooders stacked on top of each other allowing a large number of chicks to be brooded in a limited space

BEAKING OUT. A landfowl habit of using the beak to scoop feed out of a feeder onto the floor

BILLING OUT. A waterfowl habit of using the bill to scoop feed out of a feeder onto the floor

BLACKHEAD. Histomoniasis, a serious and incurable disease of young turkeys caused by the protozoan *Histomonas meleagridis,* that commonly appears toward the end of the brooding period or for several weeks thereafter

BLASTODERM. The beginning of embryo development, arising from a blastodisc that has been fertilized, at which time the cells organize into a set of clearly visible concentric rings

BLASTODISC. A pale irregularly shaped spot of cells on the surface of an egg yolk that, if fertilized with sperm, becomes a blastoderm

BLOOD FEATHERS. Pinfeathers that contain a supply of blood to nourish the feathers during their growth stage

BLOOD RING. The appearance of a candled egg in which the embryo has died early, causing blood to accumulate in a ring circling the egg's short circumference

BLOOD SPOT. A small, dark, reddish or brownish dot of blood that appears on an egg yolk, or within the white

BREAKOUT ANALYSIS. The examination of an incubated or partially incubated egg, typically to determine why the embryo failed to develop properly

BREED. A genetically pure population having a common origin, similar conformation and other identifying characteristics, and the ability to reliably produce offspring with the same conformation and characteristics

BREED TRUE. To produce offspring that are almost identical to the parents

BROILER. A young tender chicken raised for meat

BROOD. To sit on eggs until they hatch or raise a batch of hatchlings. Also the hatchlings themselves.

BROODER. A facility where hatchlings are temporarily raised without a mother hen until they have enough feathers to keep themselves warm

BROOD PATCH. A large defeathered area on a setting hen's breast

CABINET INCUBATOR. A large-capacity incubator designed to be placed on the floor

CALCITE (CALCIUM CARBONITE) CRYSTALS. The substance making up the hard portion of an egg's shell (the calcareous layer) from which a developing embryo derives calcium for bone development

CANDLING. Shining a light into an intact egg to examine its contents through the shell and determine its interior quality prior to hatching or embryo development during the hatch

CANNIBALISM. The nasty habit chicks and poults have of eating each other's feathers or flesh, which may occur when they grow their first feathers

CAROTENOIDS. Natural pigments occurring in two classes of plants: xanthophylls (consisting of leafy greens), from which egg yolks get their color, and carotenes (consisting of numerous orange fruits and vegetables). Yellow corn is the only major constituent of poultry feed that contains both xanthophyll and carotene.

CERAMIC BULB. An infrared heater made of porcelain that uses the same type of fixture as a regular screw-in light bulb but does not emit light. Also called *ceramic infrared heat emitter* and *Edison screw-base ceramic bulb.*

CHALAZAE. Fibrous cords formed by extensions of the chalaziferous layer that twist together on opposite sides of the yolk, anchoring the chalaziferous layer to the shell membrane and protecting the yolk by suspending it and centering it within the albumen

CHALAZIFEROUS LAYER. The albumen that immediately surrounds an egg's yolk, cushioning the yolk and containing defenses against bacteria. Also called *inner thick.*

CHORION. A membrane that surrounds and protects a developing embryo and its support systems

CLOACA. The chamber just inside the vent where the digestive, excretory, and reproductive tracts come together

CLOACAL SEXING. See **vent sexing**

CLUBBED DOWN. Down on a newly hatched chick, on which the sheaths fail to rupture, thus bending each bit of fluff so it bulges out at the shaft base and giving it a club-shaped appearance

CLUTCH. A batch of eggs or brood of chicks that hatch together. Also the eggs a hen produces in one laying cycle

COCCI. Short for **coccidiosis**

COCCIDIOSIS. The serious condition of being infected with coccidia, various species of parasitic protozoa; most of these belong to the genus *Eimeria*, the main signs of which are slow growth and loose, watery, or off-color droppings

COCCIDIOSTAT. One of several drugs used to inhibit the development of coccidiosis

COIR. Shredded coconut husk widely used as a peat substitute for bedding. Also called *cocopeat* and *coconut bedding.*

COLOR SEXING. Taking advantage of the sex-link gene that controls feather color to sort chicks by the color or color patterns of their down; a practice commonly used to produce hybrid brown-egg layers known as black sex links and red sex links

CONGENITAL LOCO. A spasming, epilepticlike nervous condition of chicks and poults that causes a hatchling's head to bend so far back its beak points toward the sky and the birds tips. Also called *stargazing.*

CONTAGIOUS. Description of a disease that spreads by direct or indirect contact with an infected individual

CONTINUOUS HATCHING. Periodically putting fresh eggs into an incubator so that while some hatch, others are at various stages of incubation

CREST. A spherical puff of feathers growing on the head of a breed that has a knob on top of its skull

CROP. An expandable pouch at the base of a bird's neck that bulges with feed after the bird has eaten and where digestive juices begin softening the feed before it moves into the bird's stomach

CROSSED BEAK. A beak in which the upper and lower halves grow in opposite directions, preventing the bird from pecking properly

CUCKOO. A feather color pattern consisting of coarse, irregular barring

CULL. To remove an undesirable individual (for instance, an egg, hatchling, or breeding bird) from the population at large. Also, the individual thus removed.

DEBEAKING. To shorten a beak to prevent cannibalism. Temporary debeaking may be considered for chicks that persistently peck each other and cannot be stopped.

DIGITAL EGG MONITOR. A device that uses infrared technology to detect and display on a readout screen the heart rate of an embryo inside an egg placed on a sensor pad

DIMORPHISM. Readily observable differences in the physical features and behavior patterns between males and females of the same breed

DORSAL STRIPES. Stripes that run along the back, usually on both sides of the backbone, the color and length of which sometimes can be used to color-sex chicks; also called *chipmunk stripes*

DOWN. The soft furlike fluff covering a hatchling

DRINKER. A container from which birds drink water; also called *waterer* or *fount*

DRY-BULB TEMPERATURE. The temperature of air as measured by a thermometer that is freely exposed to the air; the dry-bulb temperature in a properly maintained incubator is higher than the wet-bulb temperature

DUAL-PURPOSE BREED. Any breed developed for the production of both eggs and meat

EGG TOOTH. A small, sharp, temporary cap at the tip of a hatchling's upper beak used to break out of the shell

EGG TURNER. A device that either tilts or rolls eggs to periodically change their position during incubation

ENDEMIC. Description of a disease or pathogen that is regularly present in a particular area

EVAPORATION PAN. A shallow container in an incubator, from which water evaporates to provide humidity. Also called *water pan*

EXPLODER. A rotting egg that literally explodes in an incubator or under a hen

EXTRAEMBRYONIC MEMBRANES. Membranes that develop inside an incubated egg to help the embryo develop, live, and grow but that do not become part of the hatchling's body

FEATHER TRACT. A specific area of the skin in which feathers grow

FEEDER. A container from which birds eat their daily rations

FERTILE EGG. An egg containing sperm, making it capable of producing a baby bird when incubated under appropriate conditions

FOOD CALL. A high-pitched staccato sound repeated rapidly by a hen to bring her chicks' attention to something tasty to eat

FORCED-AIR INCUBATOR. An incubator with a built-in fan that constantly circulates air to maintain an adequate oxygen level and keep the temperature even throughout the unit. Also called *circulated-air incubator* or *fan-ventilated incubator.*

FOUNT. A *drinker*

FREE RADICALS. Highly reactive chemicals that are normal by-products of metabolism and which modify chemical structures, sometimes causing tissue damage

FREE RANGE. Technically, a description of birds that are not confined, but in practical terms a description of birds that are allowed to roam at will within a fenced outdoor area

FULL MOLT. The renewal of feathers in all of the feather tracts, as opposed to a partial molt that affects some but not all tracts

GERMINAL DISC. A pale, irregularly shaped spot of cells on the surface of an egg yolk. If fertilized with sperm, it becomes a blastoderm. Also called *blastodisc.*

GLOBULINS. Spherical-shaped, soluble, simple proteins widely found in plants and animals, including an egg's albumen

GROW UNIT. Intermediate housing for birds that have outgrown their brooder but are still too small to be safe and comfortable in facilities designed for full-grown birds. Also called *grow-off unit, grow pen,* and *halfway house.*

GROW-OFF UNIT. A grow unit, generally used for raising broilers from the time they are removed from a brooder until they are ready for butchering

GYNANDROMORPH. A bird (or other creature) that has both male and female characteristics and organs

HALFWAY HOUSE. A grow unit

HATCH. The process by which a newly developed bird escapes from within the egg shell; also a group of baby birds that hatched together (as in, "the hatch")

HATCHABILITY. The ability of fertile eggs to hatch, expressed as a percentage of those that successfully hatch in a group incubated together

HATCHER. A mechanical device similar to an incubator, but without an egg turner, into which eggs taken from an incubator are placed a couple of days before they start hatching

HATCHERY. A place that incubates and hatches eggs for commercial purposes

HATCHING EGG. A fertilized egg properly handled to best maintain its hatchability

HATCHLING. A recently hatched bird

HELP-OUT. A bird that is unable to hatch from an egg without assistance

HEN FEATHERED. A cock's plumage that is nearly identical in color and markings to a hen's of the same breed and variety

HISTOMONIASIS. An incurable disease of young turkeys caused by the protozoan *Histomonas meleagridis* that commonly appears near the end of the brooding period or for several weeks thereafter. Also called *blackhead.*

HOVER. A heat source hung above brooded chicks and designed to allow chicks to warm themselves as needed or to move away from the heat to eat, drink, and exercise

HUMIDITY PADS. Sponges used in an incubator to increase the surface area available for evaporation, thus increasing the humidity level. Also called *wick pads.*

HYBRID. A population parented by females of one breed and males of another breed — generally for the purpose of increased efficiency in egg production or rapid growth for meat production — having similar conformation and other identifying characteristics, but not the ability to reliably produce offspring with the same conformation and characteristics

HYBRID VIGOR. The phenomenon whereby cross-bred chicks are stronger and healthier than either of their parents. Hybrid vigor is the opposite of inbreeding depression.

HYGROMETER. An instrument for measuring the amount of moisture in the air in either wet-bulb degrees (as measured by a wet-bulb thermometer) or percent relative humidity (as measured by a digital hygrometer)

INBREEDING DEPRESSION. Decrease in vigor and fitness in the offspring of inbred parents

INCANDESCENT HEAT. Warmth created by a source that produces light by being heated, such as a light bulb

INFRARED HEAT. Warmth generated by electromagnetic energy that does not involve light

INTESTINAL MICROFLORA. A complex community of beneficial bacteria and yeasts that live in the digestive tract, aid digestion, and stimulate the immune system

KERATIN. Fibrous protein that forms the structural basis for feathers and claws

LAMELLAE. Tiny comb-like ridges waterfowl use like teeth to bite off bits of vegetation, to grab and hang onto insects, and to strain food particles out of water

LANDFOWL. Chickens and chicken-like birds that live primarily on the ground and have short, rounded wings suitable for short-distanced flight

LEG BAND. A method of identification consisting of a plastic or aluminum ring that wraps around a bird's leg

LETHAL GENE. A gene that can cause death, typically during incubation

MALPOSITION. Any position of a developed embryo that would prevent it from being able to free itself from the shell at the time of hatch

MAREK'S DISEASE. A fairly common poultry disease caused by a highly contagious herpes virus that results in tumors, paralysis, and suppression of the immune system

MEDICATED STARTER. A starter ration containing a coccidiostat to prevent coccidiosis

MINIATURES. Bantam breeds having a larger counterpart.

MOLT. The periodic shedding and renewal of plumage, controlled by hormones and regulated by daily exposure to light. Chicks go through a complete molt at one to six weeks of age and partial molts at seven to nine weeks, 12 to 16 weeks, and 20 to 22 weeks.

MULTI-STAGE INCUBATION. The procedure of placing groups of eggs in an incubator, typically a cabinet unit, at different times so that they hatch at different times, as opposed to single-stage incubation. Also called *continuous hatching.*

MUSHY CHICK DISEASE (OMPHALITIS). A condition in which the yolk sac isn't completely absorbed so the navel can't heal properly; as a result bacteria invade through the navel, causing chicks to die at hatching time and for up to two weeks afterward

OLIGOSACCHARIDES. The most common prebiotics consisting of non-digestible carbohydrates

PANEL HEATER. A nonlight-emitting, energy-efficient, thin, rectangular heat source that produces soft, uniform heat over a more specific area than does a heat lamp

PARTHENOGENESIS. Development of an embryo in an egg that has not been fertilized. Also called *asexual reproduction*

PARTIAL MOLT. The renewal of feathers in some, but not all, of the feather tracts

PASTING. A common condition in newly hatched chicks whereby soft droppings stick to a chick's vent, harden, and then seal the vent shut, eventually causing death. Also called *paste up*, *pasty butt*, and *sticky bottom*.

PECKING. Using the beak to bite or strike something

PECK ORDER. The social hierarchy that develops among a population of birds and determines such things as which ones eat first or roost on the highest perches

PEDIGREE. A complete record of a bird's ancestry

PEDIGREE BASKET. A basket-like enclosure used in an incubator or hatcher to separate chicks resulting from different matings

PICKING. Undesirable landfowl behavior involving pulling out each other's feathers or pecking at flesh to the point of causing open wounds

PINFEATHERS. The pinlike tips of newly emerging feathers appearing when a bird first feathers out and again whenever it undergoes a molt

PIP. The hole a newly formed chick makes in its shell when it is ready to hatch. Also the act of making such a hole

PLUMULES. Short feathers that lack barbicels, making them fluffy rather than smooth

PREBIOTICS. Non-digestible carbohydrate fibers that stimulate the growth and activity of gut flora, commonly used in conjunction with probiotics

PRECOCIAL. Capable of independent activity, including self-feeding, almost from the moment of birth; from the Latin word praecox, meaning mature before its time

PRIMARY FEATHERS. The 10 largest feathers of the wing

PROBIOTICS. Live microflora introduced as a dietary supplement that is either dissolved in water or added to feed. Their purpose is to boost immunity by giving chicks an early dose of the same gut flora that will eventually colonize their intestines.

PROLACTIN. A hormone, released by the pituitary gland when day lengths increase, that triggers broodiness

PTERYLAE. Feathered areas of the skin that are separated by featherless areas (apteria)

RANGE SHELTER. Portable housing for pasturing poultry

RHEOSTAT. A dimmer switch used to adjust light or heat output

RUMPLESS. Lacking a tailbone and therefore lacking a tail

RUN. A fenced-in outdoor area that gives poultry access to sunshine, fresh air, and exercise. Also called *pen* or *yard*.

SALMONELLA. A genus of bacteria that includes many different species, a few of which affect poultry. These bacteria occur mainly in the intestine and can cause intestinal diseases. Also an infection caused by such a bacterium.

SET. To place a group of eggs together in an incubator or under a hen. Also (said of a hen) to keep eggs warm so they will hatch. Also called *brood*.

SETTING. A group of eggs placed together in an incubator or under a hen

SEXED CHICKS. Chicks that have been sorted into two groups: males and females

SEXING. Separating the males from the females

SEX LINK. A hybrid cross, bred to take advantage of sex-linked genes so its gender at hatch may be determined by physical appearance (color sexing or feather sexing), rather than by a microscopic examination of its sex organs (vent sexing)

SEX-LINKED GENES. Genes carried on the pair of chromosomes that determine a bird's gender

SINGLE-STAGE INCUBATION. Setting eggs in an incubator so they all hatch at the same time. Also called *all-in all-out incubation*.

SPIRALS. Thin, spiral leg bands used to identify poultry by color code

STARGAZING. A nervous condition of chicks and poults that causes a hatchling's head to bend so far back its beak points toward the sky. Also called *congenital loco.*

STILL-AIR INCUBATOR. A mechanical device for hatching eggs that lacks a fan to circulate air. Also called *gravity-flow* or *gravity-ventilated incubator.*

STRAIGHTBRED. The offspring of a male and female of the same breed and variety. The term *straightbred* is more accurate than the term *purebred,* since poultry have no registry and therefore no registration papers as proof of lineage.

STRAIGHT-RUN CHICKS. See: *as-hatched chicks.*

STRAIN. A related population of poultry — which may be either straightbred or hybrid — having nearly identical conformation and other identifying characteristics that make them especially suitable for a specific purpose, such as meat production, egg production, or exhibition. Also called *line.*

STRESSOR. Any negative experience, situation, or event that causes psychological or physical tension or anxiety

SUN PORCH. An open pen that is raised above ground, designed to provide light and air under sanitary conditions

TABLETOP INCUBATOR. A small capacity incubator, usually holding 50 eggs or fewer, that is typically placed on a table or countertop

TIDBIT. To repeatedly pick up and drop a bit of food while sounding the food call. A mother hen tidbits to break up a large food item into pieces small enough for her chicks to peck.

TOE PUNCH. A device used to punch holes into the webbing of a hatchling's feet for purposes of identification.

TROUGH. A long narrow drinker or feeder

TRUE BANTAM. A small breed that lacks a larger counterpart

TUBE FEEDER. A cylindrical feeder into which rations are poured at the top and birds eat from a circular pan at the bottom

UNSEXED CHICKS. See: *as-hatched chicks.*

UTILITY BREED. A dual-purpose breed

VENT. The outer opening of the cloaca, somewhat analogous to the anus

VENT HOLES. Small holes cut into the top or sides of an incubator to release stale air and admit fresh air

VENT SEXING. A traditional Japanese method of determining a hatchling's gender by examining minor differences in the tiny cloaca just inside a vent. Also called *cloacal sexing.*

VITELLINE SAC. See: *yolk sac.*

WATERER. A container from which birds drink water. Also called *drinker* and *fount.*

WATER PAN. A shallow container in an incubator, from which water evaporates to provide humidity. Also called *evaporation pan*

WET-BULB TEMPERATURE. A measure of the amount of moisture in the air, as determined by fitting the end of a thermometer with a moistened wick from which water evaporates; the wet-bulb temperature in a properly maintained incubator is lower than the dry-bulb temperature

WICK PADS. Humidity pads

WING BAND. A method of identification consisting attaching an aluminum strip through a bird's wing web, commonly at the time of hatch

WING WEB. The triangular flap of skin between a bird's shoulder and the wing's second main joint, which stretches and becomes more easily visible when the wing is fully extended

WITHDRAWAL PERIOD. The minimum number of days that must pass from the time a bird stops receiving a drug until the drug residue remaining in its meat is reduced to an "acceptable level" established by the United States Food and Drug Administration, based on the drug manufacturer's recommendation, and considered safe for human consumption

XANTHOPHYLL. A natural, fat-soluble yellow-orange pigment found in green plants and yellow corn. It gives egg yolks a rich yellow color and colors the skin and shanks of breeds developed to have yellow skin.

YOLK SAC. A membrane surrounding the yolk where the first blood vessels form during the first day of incubation. Its function is to transport nutrients from the yolk to the embryo; as the embryo grows and the yolk gets used up, the membrane gets smaller. Also called *vitelline sac.*

ZYGOTE. A fertilized egg before embryonic cell division begins to take place. The zygote contains all the essential elements for the development of a baby bird, but they remain encoded as a set of instructions until incubation allows the zygote to develop into an embryo.

ACKNOWLEDGMENTS

Thanks go to the following people for assistance in their individual areas of expertise: David Kemp of Kemp's Koops and Incubators, Chuck and Sonja Scharf of Infratherm, Jeff Smith of Cackle Hatchery, Susan E. Watkins of the University of Arkansas Center of Excellence for Poultry Science, Shane Bagnall of Zoo Med Laboratories, Eric Stromberg of Stromberg's Chicks and Game Birds Unlimited, Ralph Winter of Guinea Farm, Willie Strickland of GQF Manufacturing, Tom Wenstrand of Brower Incubators at Hawkeye Steel Products, photo editor Mars Vilaubi of Storey Publishing, bird behavior specialist and mentor Gene Morton, awesome editor Rebekah L. Boyd-Owens, illustrious illustrator Bethany Caskey, and my husband, Allan Damerow — brooder builder extraordinaire.

RESOURCES

Hatching and Brooding General Suppliers

Avian Biotech International
Tallahassee, Florida
800-514-9672
www.avianbiotech.com/buddy.htm
Digital egg monitor

Brinsea Products, Inc.
Titusville, Florida
888-667-7009
www.brinsea.com
Incubators, brooders

Eagle Nest Poultry
Oceola, Ohio
419-562-1993
www.eaglenestpoultry.com
Hatching eggs, hatchlings

EggCartons.com
Manchaug, Massachusetts
888-852-5340
www.eggcartons.com
Incubators, brooders, general supplies

G.Q.F. Manufacturing Co.
Savannah, Georgia
912-236-0651
www.gqfmfg.com
Incubators, brooders

Hoffman Hatchery, Inc.
Gratz, Pennsylvania
717-365-3694
www.hoffmanhatchery.com
Hatchlings, incubators, brooders, general supplies

Holderread Waterfowl Farm & Preservation Center
Corvallis, Oregon
541-929-5338
www.holderreadfarm.com
Duckling and gosling specialists

Hoover's Hatchery
Rudd, Iowa
800-247-7014
www.hoovershatchery.com
Hatchlings, starter kits, general supplies

Incubators.org
Crittenden, Kentucky
800-259-9755
www.incubators.org
Incubators, brooders

Ideal Poultry Breeding Farms, Inc.
Cameron, Texas
254-697-6677
www.idealpoultry.com
Hatchlings

Infratherm, Inc.
Sarona, Wisconsin
715-469-3280
www.sweeterheater.com
Infrared panel heaters

Kemp's Incubators
WireLink USA
Eugene, Oregon
888-901-2473
www.poultrysupply.com
Incubators, brooders, general supplies

Lehman's
Kidron, Ohio
888-438-5346
www.lehmans.com
Kerosene-powered incubators

Lyon Technologies Inc.
Chula Vista, California
888-596-6872
www.lyonusa.com
Incubators, some supplies

Metzer Farms
Gonzales, California
800-424-7755
www.metzerfarms.com
Duckling and gosling specialists

Meyer Hatchery
Polk, Ohio
888-568-9755
www.meyerhatchery.com
Hatching eggs, hatchlings, incubators, brooders, general supplies

Murray McMurray Hatchery
Webster City, Iowa
800-456-3280
www.mcmurrayhatchery.com
Hatching eggs, hatchlings, incubators, brooders, general supplies

My Pet Chicken, LLC
Norwalk, Connecticut
888-460-1529
www.mypetchicken.com
Hatching eggs, chicks, incubators, brooders, general supplies

Privett Hatchery, Inc.
Portales, New Mexico
877-774-8388
www.privetthatchery.com
Hatchlings

Randall Burkey Company
Boerne, Texas
800-531-1097
www.randallburkey.com
Hatching eggs, chicks, incubators, brooders, general supplies

Ridgeway Hatchery, Inc.
LaRue, Ohio
800-323-3825
www.ridgwayhatchery.com
Hatchlings, some supplies

Sand Hill Preservation Center
Calamus, Iowa
563-246-2299
www.sandhillpreservation.com
Hatchlings

Smith Poultry & Game Bird Supply
Bucyrus, Kansas
913-879-2587
www.poultrysupplies.com
Incubators, brooders, general supplies

Stromberg's Chicks & Gamebirds Unlimited
Pine River, Minnesota
800-720-1134
www.strombergschickens.com
Hatching eggs, hatchlings, incubators, brooders, general supplies

Sun Ray Chicks Hatchery
Hazelton, Iowa
319-636-2244
www.sunrayhatchery.com
Egg and meat hatchling specialists

Surehatch
Santa Monica, California
888-350-2221
www.surehatch.com
Incubators, brooders, general supplies

Townline Hatchery
Zeeland, Michigan
888-685-0040
www.townlinehatchery.com
Hatchlings

Welp, Inc.
Bancroft, Iowa
800-458-4473
www.welphatchery.com
Hatchlings, incubators, brooders, general supplies

Zoo Med Laboratories, Inc.
San Luis Obispo, California
888-496-6633
www.zoomed.com
Heat lamps, ceramic bulbs, fixtures, safety covers

Publications

Backyard Poultry
800-551-5691
www.backyardpoultrymag.com
A bimonthly magazine devoted to all topics related to poultry, including breeder flock management, incubation, and brooding

Damerow, Gail. *Storey's Guide to Raising Chickens*, 3rd ed. Storey Publishing, 2010.
A complete guide to raising chickens, including breeder flock management, breeding plans, mating options, and progeny testing

Hunt, F. B. *Genetics of the Fowl.* Norton Creek Press, 2003.
First published in 1949, this book comprehensively covers the breeding of chickens for such things as egg production, body size, plumage patterns, and disease resistance

Jull, Morley A. *Poultry Breeding.* John Wiley & Sons, 1952.
Now out of print, this book is one of the best guides to breeding chickens for fertility and hatchability, as well as for improved meat and egg production

Lamon, Harry M., and Rob R. Slocum. *The Mating and Breeding of Poultry.* Lyons Press, 2003.
First published in 1920, this book comprehensively covers the principles of breeding chickens for improved laying and exhibition

Plamondon, Robert. *Success with Baby Chicks.* Norton Creek Press, 2003.
A guide to area-brooding sizable batches of broilers and layers for pasture production

Organizations

American Bantam Association (ABA)
www.bantamclub.com

American Livestock Breeds Conservancy (ALBC)
919-542-5704
www.albc-usa.org

American Poultry Association (APA)
724-729-3459
www.amerpoultryassn.com

National Poultry Improvement Plan (NPIP)
Animal and Plant Health Inspection Service, USDA
www.aphis.usda.gov/animal_health/animal_dis_spec/poultry

Rare Breeds Canada (RBC)
613-445-0754
www.rarebreedscanada.ca

Society for the Preservation of Poultry Antiquities (SPPA)
570-837-3157
http://sppa.webs.com

Websites

BackYard Chickens
www.backyardchickens.com

TheCityChicken.com
http://home.centurytel.net/thecitychicken

FeatherSite
www.feathersite.com

poultryOne
http://poultryone.com

Small Flock Information
University of Arkansas Cooperative Extension Service
www.aragriculture.org/poultry/small_flock_information.htm

Small Flock Management of Poultry
Mississippi State University Extension Service
http://msucares.com/poultry/management

Breed Associations

Breed Associations and Clubs
Poultry Extension, University of Kentucky
www.ca.uky.edu/smallflocks/links/breed_associations.html

Heritage Breed Poultry
Smith Poultry & Game Bird Supply
www.heritagebreedpoultry.com/heritage_breed_poultry_links.php

Cooperative Extension Service

For fact sheets on incubation and brooding, or for answers to specific questions, contact the Cooperative Extension Service in your state. This program is affiliated with each of the nation's land-grant universities and the United States Department of Agriculture in Washington, DC, and can provide detailed and accurate information. To find the nearest Extension office, contact:

National Institute of Food and Agriculture
Formerly the Cooperative State Research, Education, and Extension Service
United States Department of Agriculture
Washington, DC
202-720-4423
www.csrees.usda.gov

NOTES

NOTES

INDEX

Page numbers followed by "f" indicate photographs and illustrations. Page numbers followed by "t" indicate tables.

INTERIOR PHOTOGRAPHY CREDITS

© Adam Mastoon: 147; © Arco Martin/Getty Images: 3 right; courtesy of Brinsea Products, Inc: 47 and 133 top; © courtesy of Avitronics.co.uk: 174 © Cackle Hatchery: 85 bottom left; © Dr. Jacquie Jacob, University of Kentucky: 85 top, left, and right; © fotolincs/Alamy: 3 bottom left; © Fuse/Getty Images: 3 top left; © Gail Damerow: 9, 17 right, 20 left, 27, 36, 42, 44, 46, 52, 54, 58, 65 all except top right, 66, 69, 70, 71, 72, 75, 78, 79, 80, 81, 90, 91, 97, 103, 107, 109, 114, 116, 117, 118, 126, 130, 133 middle, 137, 142 left, 144 top four eggs, 145 all except center, 150, 152, 153, 154, 162, 163, 167, 170, 171, 173, 177, 182, 185, 201, 207, and 208; courtesy of GQF Manufacturing Company: 33, 34, 129, 132, and 133 bottom; © John Metzer/Metzer Farms: 21; © Juniors Bildarchiv GmBH/Alamy: 122; Mars Vilaubi: 5 center & right, 17 left, 55, 65 top right, 142 right, 144 bottom two eggs, 145 center, and 213; © courtesy of Lehmans.com: 159; © courtesy of Lyon Technologies, Inc.: 135 and 136; © Meyer Hatchery: 1, 5 left, 7, 11, 12, 13, 15, 16, 18, 19, 23, 24, and 240; © Natali Garyachaya/Alamy: 20 right; © Photoshot/Ernie James: 3 title; © posteriori/iStockphoto.com: 164

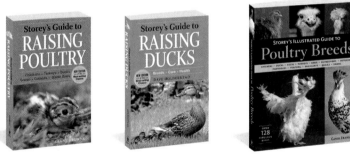